Praise for *The Jesuit and the Skull*

"The clash between science and superstition is one important theme of Amir D. Aczel's biography of Pierre Teilhard de Chardin, *The Jesuit and the Skull*. A respected paleontologist, Teilhard was a member of the team of scientists who discovered the remains of Peking Man, a promising candidate for the 'missing link' in human evolution, at Dragon Bone Hill in 1929. It was only one episode in an adventurous, tumultuous life that coincided with the wars and revolutions of the early twentieth century. A certain elegant irony lies just beneath the surface of Aczel's superb story. Teilhard took pleasure in scientific trips to Spain and France to view cave paintings—the first stirrings of religious imagination that are regarded as a line of demarcation between prehistoric hominids, essentially apes that walked upright, and the early human beings we must recognize as our direct ancestors. Tens of thousands of years later, the worst features of organized religion distorted and delimited the life and work of this visionary whom the inheritors of the Inquisition saw as a dangerous heretic. Only after Teilhard's death were his most important works printed, and only because he put the manuscripts beyond church control by bequeathing them to one of the women who had befriended him."
—*Los Angeles Times*

"With some conservative religious leaders proclaiming evolution incompatible with Christianity, the time seems to be right for a biography of Pierre Teilhard de Chardin, a Jesuit priest and scientist. Amir D. Aczel takes up the challenge in *The Jesuit and the Skull*. [It] succeeds in revealing a person for whom the evolution-creation debate wasn't a manufactured controversy designed to attract religious followers or galvanize a political base, but an attempt to reconcile faith with reason and better understand humanity's place in the universe."
—*Archaeology.com*

"An extraordinary story."
—*The Philadelphia Inquirer*

continued

"An absorbing read [and a] deeply moving personal story that sheds light on a now vanished world."

—Ian Tattersall, Curator, Division of Anthropology, American Museum of Natural History, and coauthor of *Human Origins*

"The fascinating story of the tumultuous life of Pierre Teilhard de Chardin, *The Jesuit and the Skull* is also an object lesson in how to reconcile one's religious beliefs with the study of evolution. Amir D. Aczel skillfully brings to life the struggles and achievements of this legendary scientist, and he tells a very human tale with great insight and compassion."

—Ofer Bar Yosef, Professor, Department of Anthropology, Harvard University

"I like to think of the famous Jesuit priest, Pierre Teilhard de Chardin, as a kind of clerical Indiana Jones. Ruggedly handsome, often in a bush jacket and fedora, Teilhard traveled the globe in the 1920s, searching out fossils and evidence for human evolution. Indeed, as this fine new book on Teilhard shows, the priest's work in uncovering and analyzing the remains of 'Peking Man' did indeed support evolution. It provided a key 'missing link' in primate ancestry, a tie between humans and apes. Now if that strikes you as a strange way for a Roman Catholic priest to spend his time, you would not be alone. The Vatican, for example, was never too crazy about Teilhard's work and hounded him for much of his career. Plenty of the Jesuit superiors were equally displeased. Exiled from France, forbidden to publish on evolution, denied prestigious academic appointments, Teilhard suffered mightily for his conviction that religion and science are not competitors. He saw them as friendly companions on the road to truth. Each has something to contribute to understanding the inexhaustible puzzle that the universe presents. The great merit of *The Jesuit and the Skull* is in telling the story of Teilhard's distinguished scientific career (and in his effort to integrate his views of science and his Christian faith) in a way that is accessible to readers who are neither scientists nor theologians. Amir D. Aczel so skillfully explains the basic science and is such a good storyteller, that I was caught up in the scientific discovery and taken by the personal anguish of this remarkable priest."   —*The Cleveland Plain Dealer*

"[Aczel's] narrative comes alive."   —Associated Press

"Earlier research into mathematical mysticism well qualifies Aczel for interpreting the life of Teilhard de Chardin, a cleric-scientist who defied the boundaries of both rational science and scriptural orthodoxy. Readers will marvel at how loyal Teilhard remained to a church that repeatedly disciplined him for heresy in his evolutionary explanation of human origins. It was, ironically, by exiling Teilhard from his beloved France that church authorities put him in China, where in 1929 he shared in the discovery of the famous Peking Man fossils. Aczel details Teilhard's role in that discovery, highlighting his involvement with Lucile Swan, an American artist commissioned to sculpt the ancient hominid. That relationship finally foundered when Teilhard refused to break vows of celibacy sanctified by a church that repaid his fidelity with continued hostility. Nonetheless, Aczel discerns an abiding legacy in the words and writings of a thinker who suffered much for his synthesis of pioneering science and iconoclastic faith."   —*Booklist*

"Will touch even the most obstinate soldiers of today's culture wars."

—*Seed* magazine

RIVERHEAD BOOKS

New York

THE
JESUIT
AND THE
SKULL

Teilhard de Chardin,

Evolution, and the

Search for Peking Man

•••••••••

Amir D. Aczel

RIVERHEAD BOOKS
Published by the Penguin Group
Penguin Group (USA) Inc., 375 Hudson Street, New York, New York 10014, USA •
Penguin Group (Canada), 90 Eglinton Avenue East, Suite 700, Toronto, Ontario M4P 2Y3, Canada
(a division of Pearson Penguin Canada Inc.) • Penguin Books Ltd., 80 Strand, London WC2R 0RL,
England • Penguin Group Ireland, 25 St. Stephen's Green, Dublin 2, Ireland (a division of Penguin
Books Ltd.) • Penguin Group (Australia), 250 Camberwell Road, Camberwell, Victoria 3124, Australia
(a division of Pearson Australia Group Pty. Ltd.) • Penguin Books India Pvt. Ltd., 11 Community
Centre, Panchsheel Park, New Delhi—110 017, India • Penguin Group (NZ), 67 Apollo Drive,
Rosedale, North Shore 0632, New Zealand (a division of Pearson New Zealand Ltd.) • Penguin Books
(South Africa) (Pty.) Ltd., 24 Sturdee Avenue, Rosebank, Johannesburg 2196, South Africa

Penguin Books Ltd., Registered Offices:
80 Strand, London WC2R 0RL, England

While the author has made every effort to give accurate telephone numbers and Internet addresses at
publication time, neither the publisher nor the author assumes any responsibility for errors, or for
changes that occur after publication. Further, the publisher does not have any control over and does not
assume any responsibility for author or third-party websites or their content.

First Riverhead hardcover edition: October 2007
First Riverhead trade paperback edition: November 2008
Riverhead trade paperback ISBN: 978-1-59448-335-6

The Library of Congress has catalogued the Riverhead hardcover edition as follows:
Aczel, Amir D.
The Jesuit and the skull : Teilhard de Chardin, evolution, and the search for
Peking Man / Amir D. Aczel.
p.   cm.
Includes bibliographical references and index.
ISBN 978-1-59448-956-3
1. Peking Man.   2. Fossil hominids—China.   3. China—Antiquities.   4. Evolution—
Religious aspects—Catholic Church.   5. Teilhard de Chardin, Pierre.   I. Title.
GN284.7.A29   2007   2007023581
569.9'7—dc22

PRINTED IN THE UNITED STATES OF AMERICA

10 9 8 7 6 5 4 3 2 1

For Debra, with much love

CONTENTS

THE
JESUIT
AND THE
SKULL

PROLOGUE

On a sultry day in August 2005, I traveled from Beijing to Zhoukoudian (Chou Kou Tien), a village in the hilly countryside about thirty miles southwest of the Chinese capital. I was retracing the journey made by a French Jesuit priest whose investigations of the relation between science and faith, as well as his infelicitous standing with his own church, brought him to this remote location in the Western Hills more than eighty years ago.

In 1923, Father Pierre Teilhard de Chardin arrived in China as an exile, by the order of his superiors at the Jesuit headquarters in Rome. His transgressions included questioning the doctrine of original sin and supporting Darwin's theory of evolution. Teilhard was a philosopher, theologian, geologist, and highly gifted paleontologist, and in his lifetime he made remarkable contributions to the theory of human evolution. But he paid dearly for his great achievements: Throughout his life, he was under relentless attack by the Catholic Church for his unorthodox views, he had to live for decades far from his beloved Paris, and his intermittent trips to France and the public lectures he gave while in Europe were always under tight scrutiny by the Jesuit order and the Vatican. He was continually denied permission to publish his books, and these seminal contributions to philosophy and religion appeared in print—to great acclaim—only after his death.

But despite the unusually harsh censures and the extreme intellectual and psychological hardships that he suffered, Teilhard never gave up his struggle to integrate science and religion, nor did he ever seriously consider leaving the Society of Jesus. He endured every reprimand, every punishment, every insult, and took in stride censorship, intolerance of his views by religious authorities, and years of exile with total obedience to his church.

And as fate would have it, in 1929, in China—that remote country to which the Jesuits had banished him to suppress the dissemination of his views on evolution—Teilhard became a crucial player in one of the most important discoveries in modern anthropology, and one that provided us with invaluable evidence for the theory that Charles Darwin had published exactly seventy years earlier. This great discovery was the unearthing of the skeletal remains of Peking Man. Far from making Teilhard ineffective in promoting evolution, the Church had inadvertently sent him to the one place where he could make the greatest contribution to proving the descent of man.

The momentous discovery of Peking Man took place at the site to which I was now headed, alone but for a driver who spoke no English and stared at me sympathetically as I struggled to communicate. "Why is the traffic so bad? Has there been an accident?" I asked, striking one hand against the other to indicate a car crash. The driver laughed, but said nothing, his eyes focused on the road ahead. The heat and congestion of the city made me wonder what Teilhard had felt when he arrived in China decades earlier, after a monthlong voyage from Marseille. This was well before the modern Chinese changed the Latin spelling of the name of their capital from Peking to Beijing, before construction of the two giant highways that now encircle the city, before cars had largely replaced the traditional rickshaws and bicycles, before the sky was darkened by industrial haze and pollution.

Teilhard was highly educated—with a doctorate in paleontology from the Sorbonne in addition to his religious training—spirited, worldly, and urbane. Before going to China, he had studied philosophy, literature, and science in France; taught physics in Egypt; and—because the anticlerical laws of 1901 forced the Jesuits to leave France—been ordained a priest in Hastings, England. He had survived World War I, saving lives in the trenches as a medic and stretcher-bearer in some of the most horrific battles, and at war's end was awarded a medal for his bravery by the French government.

Still, it was something of a shock for him to arrive alone in China in 1923, in this isolated, far-off land. Only the coastal Chinese were accustomed to Western faces, and I wondered how Teilhard coped with the enormous cultural differences between China and the world he knew. I recalled one of the yellowing photographs of him I had seen, taken sometime in the late 1920s, showing him standing among a group of Chinese and Western colleagues, dapper in a neatly pressed khaki field jacket—not the black Jesuit cassock, which he wore only occasionally in Asia. He looked content, confident—perhaps even relieved to be away from Europe and the controversy that surrounded him wherever he went on that continent.

Finally having inched our way out of Beijing, we found ourselves driving southwest on a two-lane country highway. We passed wide, cultivated fields and factories with billowing smokestacks. This was modern China: the ancient and the new side by side, without a wasted square inch. The driver pressed hard on the gas, pulling up directly behind the car in front of us until the driver changed lanes. My back ached from the stop-and-go traffic, but there was no use fumbling for the seat belt—they do not exist in Chinese cabs.

After an hour, we turned off the highway and onto a bumpy dirt road. This felt worlds away from the dazzling modernity of Beijing. As we passed increasingly rural villages, I saw fewer cars, and the people rode bicycles and wore simple clothing. At the road's end, the driver stopped, opened the door, and said, "Zhoukoudian." He lit a cigarette while waiting for me to come out. My translator back at the hotel had assured me the driver would wait for me here.

I emerged to see a green hill, some two hundred feet high, steep and densely overgrown with vegetation. This was the legendary Dragon Bone Hill. It was named so for its abundant supply of fossilized animal remains —"dragon bones" to the rural Chinese. The hills here are made up of limestone and chalk, a geological formation especially favorable to the preservation of skeletons of prehistoric animals—and remains of human ancestors. Nevertheless, local people still believe it to be a mystical spot where dragons come to die; some of the more superstitious refuse to venture up the hill at night.

By day, however, until the government banned the practice decades ago, people dug up the hill with shovels and pickaxes in search of bones to sell on the medicinal market. Dragons have long been a symbol of strength and vitality in Chinese culture, and their bones were said to hold great curative powers and were valuable commodities. Chinese apothecaries often paid substantial sums for these "dragon bones," which were then ground into fine powders and sold as cures for a variety of ailments, from skin rashes to insomnia to impotence.

Dragon Bone Hill is now an unassuming spot, and, I gathered, few tourists had ventured here in recent years. The fresh forest air, smelling of pine, was invigorating, and I took a deep breath and looked around. The hill I was facing was situated northwest of the village of Zhoukoudian, across the Zhoukou River from me. But it was so quiet and green here that it seemed miles from civilization. It was late afternoon, and I could see the sun slowly descending as I faced the hill. I shouldered my backpack and started from the parking area on a path leading straight up, surrounded by a forest of small trees and shrubs. It was eerily quiet. There was not a person in sight; even my driver had disappeared. As I climbed higher, a chorus of crickets and birds greeted me and the trees closed around me.

Halfway up the hill, I encountered an old sign in Chinese and English, directing me to the "Peking Man Site." I turned and worked my way down, climbing over fallen tree trunks, thick bushes, and ferns. The sound of the crickets was now nearly deafening. Suddenly I reached a cave, and a set of broken stairs leading to its bottom. I cautiously edged my way down the steps, gripping the limestone walls for balance. I was now inside a cave located halfway up the hill. The roof of the cave was open on two sides, and through these openings the sun's rays entered, providing enough natural light by which to see. It took me a few moments to orient myself. And then I saw it. Among the ferns and seedlings that seemed to grow out of the sheer rock face, I could make out faint markings in Chinese characters. From the chart and the old photographs I had brought with me, I recognized the rock surface of the cave wall. I was now standing at the exact spot where, on a snowy December 2, 1929, just as the last rays of the setting sun were barely warming the almost frozen earth, a group of diggers led by the young Chinese paleontologist Pei Wenzhong pulled out the first skull of *Homo erectus* from the rubble at the farthest corner of the cave. This was the fossilized skull of Peking Man, still ensconced in thick clay.

A few years earlier, a high-powered international team of scientists, including prominent anthropologists, geologists, paleontologists, and anatomists, had begun to come together at Dragon Bone Hill with the purpose of exploring for fossils of human ancestors. This group ultimately included the Swedish geologist Johan Gunnar Andersson, the renowned German Jewish doctor and anthropologist Franz Weidenreich, the tireless and gifted Canadian anatomist Davidson Black, and the French cleric Pierre Teilhard de Chardin.

It took more than eight years of intensive, sometimes discouraging work to uncover the first skull. Pictures of Dragon Bone Hill taken in the 1920s show a sprawling, dusty excavation complex, empty of trees and plants, extensively dug up to uncover ancient fossils. (A fossil is what a bone becomes after the organic matter inside it gives way to rocklike matter, over the course of many thousands of years.)

The first skull was the pinnacle of success of this large-scale project. Within a year of the discovery, Teilhard and his colleagues proved that human ancestors had lived here, and demonstrated that the cave's inhabitants had made tools and fire: In short, these inhabitants appeared to be the link between man and ape, able to fashion tools from rocks, hunt with these implements, cook their food, and master their living environment by using fire for heat. The years that followed saw a wealth of further discoveries of skulls, bones, and tools, until the fragmentary remains of a prehistoric community of forty individuals were assembled and studied. The relics of Peking Man, the hominid (member of the family Hominidae, of humans and their ancestors) that this collection of fossils was said to represent, were an invaluable boon to science. Analysis of them provided strong evidence for evolution, and Peking Man emerged as a key example of *Homo erectus* (Upright-Walking Man)—the "missing link" between humans and apes.

But the Society of Jesus, a historically powerful Catholic order founded in the sixteenth century by Saint Ignatius of Loyola (1491–1556) and run as a semi-military religious organization (its head is designated "superior general"), was not prepared to give up the biblical story of man's creation. When the Jesuits learned of Teilhard's involvement in the discovery of Peking Man, their assault on him, a loyal member of their order, grew stronger and more vitriolic. Every word he wrote or said—publicly or even in private—was now suspect and subject to careful examination.

In part to bridge the gap between science and religion, Teilhard proposed his own theory of evolution, encompassing elements of faith, physics, and anthropology. It was an amalgam of science and religious belief that attempted to unite the two disciplines in one logical, rational whole. Teilhard held that human evolution was a scientifically proven process that agreed with scripture and brought us closer to

God as we evolve further as a species. He hoped that the Society of Jesus would tolerate this view of evolution and allow him to lecture about it and publish. But Teilhard's books were censored by the Jesuits, and permission to publish was repeatedly denied.

His work on Peking Man exemplified science at its best. For until the discovery of Peking Man, a hominid that lived about 500,000 years ago, anthropologists had only the findings made by the Dutch anatomist Eugène Dubois in Indonesia in 1891 (a set of fossilized remains of a hominid called Java Man that dated to 700,000 years ago) to support their thesis of evolution. And there were also fossils, discovered earlier in the nineteenth century, of the much later Neanderthals, who lived from 300,000 years ago or earlier until their abrupt disappearance about 30,000 years ago with the advent of the Cro-Magnons (termed "anatomically modern humans" because their skeletal remains are indistinguishable from ours). Java Man had been hailed, at first, as the missing link between humans and apes. But there was no evidence that Java Man could make fire, the quality of the finds from Java was spottier, the fossils were not as well preserved, and they represented far fewer individuals than those eventually found at the Peking Man site.

The find at Dragon Bone Hill therefore caused an immediate sensation around the world, dominating newspaper headlines and radio programs. For a brief moment, the world's attention was directed at the most conclusive evidence so far in support of human ancestry in agreement with Darwin's theory of evolution.

And then, seemingly overnight, everything was lost. The Japanese, who had occupied parts of China since the beginning of the Sino-Japanese War in 1937, consolidated their takeover and extended it over northern China once World War II broke out in 1939. After this wider occupation of China, and in an effort to keep them from enemy hands, the Peking Man fossils were placed in a sealed room at Peking Union Medical College. But two years later, in 1941, just as the United States was about to enter World War II and the conflict with the Japanese was intensifying, Chinese authorities feared that the Japanese might discover the valuable fossils and remove them to Japan.

The Chinese, along with officials of the American embassy, arranged to transfer the entire collection of Peking Man remains to safety in the United States. The finds were clandestinely packed in two crates, and readied for shipment by sea to the American Museum of Natural History in New York, to be kept there until the end of the war. But when the crates left Peking Union Medical College to be brought to the ship that was to take them to America, they vanished without a trace. Not a shred of evidence has surfaced over the intervening six and a half decades to point to what actually happened to this collection. The fate of the Peking Man fossils remains a mystery to this day, and their disappearance an incalculable loss to science.

The following chapters tell about the fantastic discovery of Peking Man. They describe how a determined group of the world's greatest anthropologists, geologists, anatomists, and other scientists resolved, against all odds, to find the missing link between humans and apes. This is the story of how these researchers provided the key piece of evidence for Darwin's theory of evolution. Their breakthrough ushered in an age of great discoveries in anthropology, with every decade bringing us closer to the complete tale of human ancestry. This is also the story of Pierre Teilhard de Chardin, an influential member of the international group of experts assembled in 1920s Peking, who courageously fought against the entrenched beliefs and doctrines of his own church in an effort to reconcile scripture with science and offer an understanding of who we are, where we came from, and where we are going. But the struggle he began more than eight decades ago continues with renewed force today.

I had not realized just how virulently the conflict between Teilhard and his church still raged—half a century after his death—until I went to Rome to research this book. Father Thomas K. Reddy, the director of the Roman Archives of the Society of Jesus at the Borgo Santo Spirito enclave, just outside the Vatican, saw me in his office on June 27, 2006. This visit had been arranged months in advance, and I had been led to believe that I would be allowed to see any document about Teilhard that I desired.

At the end of our interview, Father Reddy said that he had some particular information on Teilhard, and I asked whether I might see it.

"No," he responded, "it is confidential." I drew a sharp breath in surprise, and he added, "But you can see other things. . . ."

Pondering what I had just heard, I proceeded to the archive reading room, and ordered the first Teilhard item from the catalogue. A short time later, a dusty pile of documents, bound with faded string, was placed on my table. I untied the knot—clearly no one had looked at this collection in many years—and began to examine the contents. These were Teilhard's manuscripts, which I knew had been typed in China in the 1930s by his intimate friend the American sculptress Lucile Swan, and which he had sent to Rome in hope of gaining the Jesuits' approval to publish. As I lifted the pile of manuscripts, what looked like a folded letter of several pages fell out.

I picked it up, opened it, and studied the yellowing sheets. What I held in my hands was a curious ten-page document, carefully hand-written in Latin, and dated March 23, 1944. I was engrossed in reading it, when I suddenly looked up to see Father Reddy standing right in front of my table. "What is that?" he demanded. "What is the date?"

I told him.

He turned pale and said: "This is exactly what I didn't want you to see."

In his frustration that I had now seen this document, Father Reddy decided to seek an immediate meeting with the Jesuit superior general, Peter-Hans Kolvenbach, to discuss what could be done about my discovery. The Jesuit headquarters, the Curia Generalizia, was next door, and as Reddy hurriedly left the room, he turned to me and said: "You are a writer: Be careful with what you write. *Don't* get us in trouble with the Vatican."

Chapter 1

THE BANQUET

Science works in mysterious ways. A scientist makes a discovery unexpectedly—without prior knowledge, and often without even a guess, about what might happen once the experiment is performed or the analysis completed. When Galileo first aimed his telescope at Jupiter that starry night of January 7, 1610, he had no idea that he was about to discover "his" moons—the Galilean satellites orbiting the giant planet—and thus provide us with the first astronomical proof that Earth was not the center of the universe. Wilhelm Conrad Röntgen didn't know, on November 8, 1895, that the electrical glass tube he was testing would cause a faint glow several feet away, and that he would thus discover X-rays and open a new era in medicine. And Louis Pasteur did not imagine while conducting his experiments with bacteria in April 1862 that most of the milk we drink today would be pasteurized—treated against germs with the process he was about to discover.

On October 22, 1926, a most unusual meeting of scientists took place in the Swedish embassy in Peking, a stately building with a severe marble façade. The event they were attending was a banquet held in honor of Crown Prince Gustaf Adolf and Princess Margareta of Sweden, who were visiting Peking as part of a world tour. The scientists gathered there were all confident that they knew the future outcome of their research project. In fact, they had already named their find—and even begun to write a paper about their expected results for the prestigious journal *Nature*. They had named their anticipated discovery *Sinanthropus pekinensis*—Chinese Man of Peking. And all they had going for them thus far as an indication of what might lie ahead was a pair of tiny teeth. This was undoubtedly one of the most audacious acts of scientific hubris in history.

These scientists, who had been arriving in Peking from several countries over the previous years, formed a tight-knit group with an ambitious aim: nothing less than uncovering man's ancestry. Now they felt that they knew precisely what to do to achieve their goal. All they needed was more money; with sufficient financial support, they could make their dream of discovering *Sinanthropus* a reality. For this, they needed the pockets of a prince.

Excavating for fossils is a complex undertaking: It demands hiring large crews of diggers, procuring and detonating explosives to clear rocks, and paying for machines to move and sift through debris. Equally, expertise in many fields is needed: paleontology to recognize and categorize fossils; anthropology and archaeology to analyze elements of human activity; anatomy to know how bones relate to one another within a skeleton; biology to understand the living creatures the fossils represent; and geology to recognize the strata of rocks and date them in order to estimate the age of the fossils. The international team of scientists gathered in China already represented expertise in several of these fields—but still others had to be recruited, and all had to be financially supported so they could continue to work on this important project.

But Prince Gustaf was not just another leisurely royal—he was also an avid amateur scientist with a strong interest in anthropology and archaeology. The organizers of this elaborate banquet knew this, and they were ready to appeal to their royal visitor for help in their scientific enterprise.

In the festivities that followed the dinner, the prince and princess stood like porcelain figurines, greeting the guests at the top of a marble staircase, and then joined them to dance to the music of a military band and toast with rose-scented wine served by white-robed, gold-sashed Chinese servants. But the milieu of this visit was surreal—it was a China never seen before, and one that was never to be seen again.

By the early 1920s, China had become a republic, and the boy emperor Puyi, without power, was a virtual prisoner in the Forbidden City while regional warlords battled for political control throughout the land. This was a time of a changing attitude toward the West that was at once open and hostile. The threat of civil war loomed, and when the emperor fled the Forbidden City, the rebel Sun Yat-sen began his attacks for control of the country. In the midst of this confusion, in 1925, Sun Yat-sen died, and China was plunged even deeper into turmoil. There were now more than five million Chinese men under arms: some were members of organized armies fighting for control of the huge country, while others belonged to lawless bands of mercenaries and frequently shifted alliances, marauding and pillaging the countryside. Despite this turmoil, Peking was surprisingly calm.

The many Westerners who lived there passed their days in a kind of fantasy. They resided in luxurious palaces, centrally located south of the Forbidden City, paying ridiculously low rents and enjoying comforts beyond imagination—with staffs of servants and caretakers. They spent their time indulging in entertainments, social events, and gossip. But the people who gathered at the Swedish embassy were not typical Westerners living in Peking. They were scientists: geologists, paleontologists, biologists, and anatomists. And the visiting prince was the protector of the Swedish Research Committee—the body that controlled all Swedish funding for fossil research outside Sweden. Prince Gustaf had an appreciation for digging for ancient fossils, and was thus the perfect person to approach for funding of a major paleontological research project—one that history would judge as destined to change the way we view ourselves, our past, and our place in the universe.

THE SEARCH FOR human ancestors began in the wake of the emergence of a new field of science: paleoanthropology—the branch of anthropology dealing with fossil hominids—which was driven by discoveries of the first hominid fossils in Europe in the nineteenth century. These were the remains of Neanderthals, who were Ice Age hominids, and Cro-Magnons, hominids whose anatomy—as inferred from their fossils— was identical to that of modern humans, and who for a time coexisted with Neanderthals. Still in the nineteenth century, Eugène Dubois discovered Java Man, whose remains were much older than the Neanderthal and Cro-Magnons finds. Yet no one then would have guessed what lay buried in a nondescript hill southwest of Peking.

The story of paleoanthropology in China began in 1899, when a young German physician and naturalist, K. A. Haberer, first heard about strange bones that had been surfacing for decades in the limestone caves near Peking. One spot in particular, a hill near the village of Zhoukoudian, was rumored to be among the largest repositories of these ancient bone fossils, which Chinese legend held were dragon remains and which sold for substantial prices on the medicinal market.

In the fall of 1900, Haberer traveled to Peking, intent on seeing these dragon bones for himself. He set out for the hilly region outside the capital and joined a band of Chinese bone hunters. This was a dangerous and uncertain time for foreigners in China; in 1899 and 1900, during the Boxer Rebellion, the Chinese revolted against foreign powers meddling in their affairs, including Japan, Russia, Germany, the United Kingdom, the United States, Italy, France, and Austria-

Hungary. The violent uprisings forced Haberer to confine his search to the area around Peking. Nevertheless, within a few months, he had collected an impressive trove of more than a hundred bones of various kinds. The find that intrigued him most was a tiny tooth. It looked uncannily similar to the tooth of a human being.

In 1903, Haberer returned to Germany and showed his collection to the zoologist Max Schlosser, who determined that most of the bones belonged to various extinct mammals—giant pigs, giant elk, cave bears. Again the tiny tooth was the one item to get special attention. Schlosser surmised that it was very old, and that it had belonged to a human ancestor. He published his findings in the journal of the Bavarian Academy of Sciences, thus arousing the scientific community's curiosity about what other important finds might await discovery in those faraway caves. But because of China's continued political instability, further research was halted for more than a decade.

Then, in the summer of 1914, Johan Gunnar Andersson was hired by the Chinese imperial government to locate new deposits of metal-bearing ores, oil and gas, and other valuable natural resources. He had recently taken an interest in the young field of paleoanthropology and had come across Schlosser's paper of a decade earlier and determined to seek out the exact location of Haberer's "dragon bones." After several years of exploring the countryside, Andersson's search led him to Dragon Bone Hill, where Haberer's tooth had been discovered.

Eager to pick up where Haberer had left off, Andersson obtained funding from the Swedish royal family and began to dig in the Zhoukoudian caves. He also hired an assistant, Otto Zdansky, an ambitious Austrian scientist who had just received his Ph.D. in paleontology from a Swedish university and was eager to continue his work on prehistoric animals.

Their partnership was ill-fated from the start. While Andersson had made it plain that he was interested only in hominid remains, Zdansky focused his search on animal fossils, which he could use in his own work. He became quite successful; years later, a dinosaur he discovered at another site in China was named after him. But at Dragon Bone Hill, tensions between the two men mounted. Andersson pressed and cajoled his willful apprentice; one day he even pointed to the very cave wall where Peking Man would be found years later, and cried, "You will find the fossils here!" Zdansky ignored him.

The project was nearly out of funding the day Zdansky uncovered a humanlike tooth. Later, he found a second tooth. Such finds undoubtedly would have given Andersson the evidence he needed to go to his patrons for renewed funding. But Zdansky, intent on finishing his own research on prehistoric animals, kept the discoveries from Andersson. He even threw away sharp flakes of quartz that he also discovered— prehistoric tools, as we now know. Soon afterward, Andersson's money ran out, and Zdansky—having completed his work—returned to Sweden. Without telling anyone, he took his finds with him, including the two teeth.

In the following years, Andersson's career languished while Zdansky's thrived. Andersson was unable to achieve his goal of finding hominid fossils, and Zdansky, whom Andersson had hired for this purpose, was able to parlay his Chinese discoveries into important academic papers. But perhaps the guilt of having remained silent about the teeth weighed on him, for in 1925, four years after he had published the results of his work at Zhoukoudian to some acclaim, Zdansky confessed the story to one of his former professors in Stockholm. The professor promptly wrote Andersson, enclosing with his letter a photograph that Zdansky had taken of the two teeth.

It is not hard to imagine how Andersson must have felt when he read the letter. Yet by all accounts he was so elated to learn that his efforts had not been in vain that he never confronted Zdansky. Now he was certain he could obtain the necessary funding to resume the excavations at Zhoukoudian. As far as he was concerned, better late than never.

❖

IT WAS JOHAN GUNNAR ANDERSSON who conceived of the meeting in Peking with his native country's crown prince. Over the previous years, he had gradually been assembling his group of experts in fields related to his work. He tried to interest them in his search, despite his lack of concrete findings, and begged or otherwise lured them to join what he believed was an immensely important project. These scientists, at least loosely associated with his Herculean yet so far fruitless effort, were present at the banquet, meeting the crown prince and princess.

Among these scientists was Davidson Black, who had an M.D. from the University of Toronto and in 1919 had been appointed professor at Peking Union Medical College—an American-funded school whose main support came from the Rockefeller Foundation. Before going to China, Black had conducted research on evolution at a laboratory in Manchester, England, analyzing the fossils of the Piltdown find in Britain—which, forty years later, would be shown to have been a hoax. Since much excitement had been generated about this "discovery" and about the search for human origins, Black became enthusiastic about fossil hunting and wanted to find his own missing link. He believed that China would be the place for such a search, and when the opportunity arose to take up a position in Peking, he jumped at the chance.

Black worked very hard at the college, teaching and researching human anatomy. He had a great appreciation for China and its culture, and he liked his Chinese colleagues, many of whom became close friends. He was also very friendly with the other foreign scientists who were gathering in Peking as if summoned for what would be one of the greatest paleoanthropological discoveries of the century. Black had a busy schedule teaching anatomy and running the department, but at night, while everyone else was asleep, he pursued his passion: studying paleoanthropology.

To get his appointment, Black had applied to the Rockefeller Foundation for a research fellowship that would enable him to move to and work in Peking. Since the appointment and his grant were strictly in the field of anatomy, he had to confine his work on fossils to after hours.

Black procured cadavers for research, obtained from the Peking police department. These cadavers were mostly of people who had been executed for various crimes; the police regularly sent Black truckloads of the bodies of these executed convicts. Execution in China was by beheading, and thus the cadavers Black received lacked heads and had mutilated necks. After some time, he asked the police whether there was any possibility of getting better dead bodies for research—corpses that were intact. The next day, he received a shipment of convicts, all chained together, with a note from the police asking him to kill them in any way he chose. Horrified, Black sent the prisoners back to the police, and thereafter obtained all his cadavers from the city morgue.

Black did exceptional work in anatomy, and his department thrived. But at night, he pored over ancient bone fossils that had been discovered in eastern China. Although he was pursuing two career paths, he made sure that he had time for family and friends. He was an outgoing man, and he and his wife were involved in many social activities of the Western expatriate community in Peking. Among his close friends who were doing work in paleontology was Weng Wenhao, the director of the China Geological Survey. Through this friendship, Black started to work in earnest on looking for fossil hominids—his passion, and the personal reason that had brought him to China in the first place.

Though he confined his work on fossils to late-night sessions in his lab—slowly burning himself out with extreme overwork, for which he would eventually pay with his life—people at the Rockefeller Foundation were furious with him. His assignment was to build a good anatomy department at the college the Foundation was supporting, not to search for human ancestors. After a visit to China, a representative of the Foundation reportedly wrote him: "For the next two years at least, give your entire attention to anatomy. Perhaps by that time you will, with your young son, have other interests which will appear more important than expeditions to mythological caves."

People at the Foundation even went out of their way and withdrew funding for a series of talks on fossils by an anthropologist from the Smithsonian Institution when they learned that Black and his colleagues had requested these talks. They were concerned that the lectures might encourage Black to stray further from the goal the Foundation had set for him.

But Black was a natural fossil-hunter, and he soon made friends with others, Chinese and Westerners, who shared his passion for the search for human origins. One of these was Pei Wenzhong, a young paleontology student who would play a major role in the drama about to unfold. Another person Black met was Johan Gunnar Andersson, the Swedish geologist and avid fossil-hunter. In 1921, Black began to collaborate with Andersson in studies of remains that had just been found in Neolithic caves in Manchuria.

The pair discovered human fossil remains, dated to about 10,000 years ago, belonging to as many as forty-five individuals. This discovery indicated that China had been inhabited by early anatomically modern humans, similar to and contemporary with the Cro-Magnons of Europe. This important finding, which showed that early humans had inhabited not only Europe but also China, energized the two men and their research assistants to look for more ancient fossils, which they felt might be discovered in Zhoukoudian—the richest site for animal fossils. To embark on a serious effort to dig for such fossils demanded funding. Black, with his social skills and friendly personality, was an asset to Andersson in approaching the Swedish prince.

To this end, Andersson asked Black to prepare a strong presentation for the visiting prince, since all that had been found so far were the two human-like teeth at Zhoukoudian. Davidson Black was such an effective speaker, with his infectious smile and good humor, as well as his penetrating knowledge of anatomy and evolution, that he was able to produce a very convincing report, which he now presented to Prince Gustaf. Black called the discovery of these two teeth "a dramatic confirmation that China would be the place to give the world evidence of Pleistocene man [the Pleistocene is a geological era stretching from 1.6 million to 11,500 years ago] to go along with the earlier discovery in Java." He concluded with an optimistic prediction: "The Chou Kou Tien [Zhoukoudian] discovery therefore furnishes one more link in the already strong chain of evidence supporting the hypothesis of the Central Asiatic origin of the Hominidae." Andersson followed up Black's enthusiastic speech by adding that he was convinced that what would be found at Zhoukoudian would constitute the most important achievement of Swedish archaeology in China.

Soon after these presentations were made at the Swedish embassy, a striking figure entered the hall. All eyes turned toward him. People recognized the renowned French paleontologist and Jesuit priest Pierre Teilhard de Chardin, who had been invited to meet the prince. Tall and handsome, in an impeccably tailored clerical suit, and exuding confidence and an air of worldliness, Teilhard walked across the floor to Andersson with his hands extended forward in greeting. Andersson proceeded to introduce him to the visiting royalty. Harry Shapiro of the American Museum of Natural History remembered thinking that "his lean, bony face with its beak-like nose was like the carving of a medieval knight that one might see on the sarcophagi in ancient French churches. Even his manner had an unusual charm."

Prince Gustaf was familiar with the paleontological work of this famous Jesuit, and his presence gave the proposed enterprise some promise. Teilhard had studied paleontology under one of the towering figures of French science, Marcellin Boule, of the Museum of Natural History in Paris, and had made important discoveries on his own, examining the fossils of many mammals, and published articles and lectured widely. But securing Teilhard's cooperation had not been easy for Andersson. Teilhard had been skeptical about the significance of the two teeth that Zdansky had found, and had even written a letter to Davidson Black expressing doubts about their origin: He thought that they could very well have belonged to animals, not to any creature resembling a human. But after studying photographs of the teeth in more detail, comparing them with teeth of apes and other mammals, and discussing their morphology with Andersson and Black, he came to accept that they might belong to a human ancestor. He became enthusiastic about the project.

The crown prince listened carefully as Teilhard described his view of the exciting possibility of finding fossils of human ancestors at Dragon Bone Hill. In part because of Teilhard's support, the prince decided that very night to fund Andersson's project. The next day, a Peking newspaper published an article on the reception, and used the name "Peking Man" for the first time to describe the yet undiscovered owner of the mysterious pair of teeth.

With his funding secure, Andersson continued to assemble his team and prepared to resume the search for hominid fossils at Zhoukoudian. So far, the team included Teilhard, Black, and a paleontologist Andersson had brought over from Sweden, Birger Bohlin. And there were a number of local Chinese scientists as well. Later, the German physician Franz Weidenreich and the Scottish geologist George Barbour would join the team.

While other Western team members had gone to China lured by the promise of a great discovery, few knew that Teilhard—as accomplished a paleontologist as he was—had not gone there of his own free will. He would just as happily have stayed in his native France, close to his large family, to which he was strongly attached, and studied fossils discovered in Europe. But Teilhard was an outcast. He had antagonized the Vatican by making scientific statements that were contrary to Church dogma: He did not believe in the traditional understanding of original sin or the literal interpretation that Adam was "the first man," and he embraced Darwin's theory of evolution. In particular, a private paper he had delivered to a clerical group in Belgium in 1922 and further developed in Paris, in which he questioned the Catholic Church's doctrine of original sin, found its way to Rome and so angered Jesuit authorities that he was sent away to China. So he was now at the right place and time to follow his calling as a paleontologist, and leave his mark on one of the greatest discoveries of human evolution.

Though he was a priest and followed the traditions and teachings of the Catholic Church, Teilhard was very much a man of his time. He knew about the great advances in science that had been taking place during his life and earlier, especially in paleontology, biology, and the study of human evolution. He was no stranger to physics, either, and had studied thermodynamics and relativity. As a Frenchman, he was particularly aware of the pre-Darwinian work on evolution undertaken in France by Georges Cuvier and Jean-Baptiste de Lamarck. He was fully versed in Darwin's theory and achievements, as well as the great debate on evolution that followed his work. And Teilhard's own mentor in Paris, Boule, was the man who first studied in detail the fossils of the Neanderthals and, with his reconstruction of what these "cavemen" might have looked like, made Neanderthal a household name.

Teilhard was charismatic and mysterious. He was a priest who made lifelong friendships in the secular world, a loyal Jesuit who cared passionately about science, and a Frenchman with deep roots in his country who spent most of his life in exile. Throughout his life, he tenaciously pursued his goal of uniting science with faith and, like Galileo three centuries earlier, fought for science against formidable odds.

Chapter 2

PRELUDE TO EVOLUTION

Teilhard's work on evolution was based on a foundation in taxonomy laid two centuries before his time by Carl Linnaeus and greatly extended into the theory of evolution by Charles Darwin a century later. Teilhard's thinking about the world around him—about life, nature, and change—was profoundly influenced by the work of these two men, as well as that of French intellectuals and scientists who lived during the eighteenth and nineteenth centuries and in his own time.

The gateway to Darwin's theory of evolution was opened in the eighteenth century with the work of the Swedish physician, botanist, and zoologist Carl Linnaeus (1707–1778). He is sometimes called Carl Linné, Carl von Linné (the name he was given when King Adolf Fredrik made him a nobleman), or Carolus Linnaeus (his Latin name). Linnaeus was born on a farm in the parish of Stenbrohult, in southern Sweden, and like his father, Nils, and his maternal grandfather was expected to follow a career as a churchman. But he was interested in natural science and hoped to become a scientist instead. Since he did not excel at the school he attended in the town of Växjö, his father urged him to become a cobbler. A friend of the family saved him; having noticed that the boy was a good botanist, he recommended to his father that Carl go to university. Linnaeus was sent to the University of

Lund, and after a year transferred to the University of Uppsala, the best in Scandinavia at the time.

While a student, Linnaeus wrote a paper about flower stamens and pistils that earned him some attention and led to an offer of a part-time position at the Uppsala botanical gardens. He excelled in his botany studies at the university, and in 1732 the Royal Swedish Academy of Sciences offered to finance a trip he wanted to make to study the flora of Lapland.

Five years later, Linnaeus published a treatise based on his research on the plants of Lapland, *Flora Lapponica*. Now in his twenties, he continued his work and began to make important advances in the understanding of species. His ambitious goal was to classify all living things. In 1735, Linnaeus moved to Holland, and there earned his medical degree from the University of Harderwijk. That same year he introduced his work of genius: the construction of a system for naming and classifying every organism on the planet. He named his work *Systema Naturae*.

By the time his guidebook to nature was in its tenth edition some two decades later, Linnaeus had classified more than 7,700 species of plants and 4,400 animal species. This work popularized and made excellent use of the binomial system, which had been invented more than a century earlier by the Swiss botanist Gaspard Bauhin, but had not been applied extensively. Until Linnaeus's guide was published, biologists were using long names, with four or more words in the name of each animal or plant. The binomial system relied on the genus and the species terms only, and it proved amazingly practical in defining an unprecedented number of species. Linnaeus's system was based on observing organisms carefully and grouping them into categories according to their characteristics. A species was defined as a group of individuals that were so similar to one another that they could breed and produce fertile offspring.

There are seven major categories in this system. They are, from largest to smallest, kingdom, phylum, class, order, family, genus, and species. According to the system, different species are grouped into genera (the plural of *genus*), genera into families, families into orders, and so on up to kingdoms. The wolf, for example, belongs to the kingdom Animalia, phylum Chordata, class Mammalia, order Carnivora, family Canidae, genus *Canis*, and species *Canis lupus*. This last pair of Latin words constitutes the wolf's binomial. The dog is *Canis familiaris*. There are also subcategories within this system, such as subspecies or subfamilies. *Terrapene carolina triungui* is a subspecies of the common eastern box turtle, which is called *Terrapene carolina*; the subspecies designation *triungui* distinguishes it as having three, rather than the usual four, toes on its hind feet. The human family is Hominidae, our genus is *Homo*, and our species is *Homo sapiens*.

Linnaeus changed the way we think about the natural world, by showing that animals and plants belonged to families that shared common general features, and yet were specifically diverse in other ways. His system of classification gives us the opportunity to understand both the distinctions and the relationships among all living creatures.

The ingenious beauty of Linnaeus's scheme lies in its simplicity. Gone are the complicated names used by biologists in earlier times. The concision of his naming system helps us to comprehend the living world: Every living thing is related to other living things through this scheme of classification. Linnaeus's system inspired scientists and laypeople alike. Scientists could now look at the world with a new perspective, and the educated throughout Europe now became more interested in the natural world.

As the system gained use and applicability, a new group arose among the educated: amateur natural scientists. These explorers scoured the countryside looking for exemplars of new species—some in hopes that the new species would be named after them—and they sent their specimens to the Swedish scientist. But a major problem emerged.

It has always been clear that humans are part of the living world. Yet how do we classify ourselves? In which group do we belong? In other words: To which animals are we humans sufficiently similar so that *Homo sapiens* can be classed within the same grouping? Although the discovery of such creatures was well in the future, Linnaeus had presciently named hypothetical early forms of humans. He had two human species within his classification scheme: *Homo sapiens* (Man the Thinker) and *Homo troglodytes* (Cave-Dwelling Man).

Linnaeus placed humans within the order of primates, and it immediately got him into trouble with religious authorities. This happened in eighteenth-century Europe, even though as early as 1698 the English anatomist Edward Tyson had dissected a chimpanzee and the following year published a book that showed that apes and humans shared more attributes than apes shared with monkeys, especially in brain structure. The mere observation that humans bore a similarity to apes sparked discontent and controversy.

From the pope to the common person, people did not want to hear that they shared characteristics with apes. Catholics were especially disturbed by Linnaeus's ideas, and the pope forbade the introduction of Linnaeus's works into the Vatican. It took many years until it became legitimate to teach his ideas in Rome.

The Lutheran archbishop of Uppsala was also upset and accused Linnaeus of impiety. Thus even pre-Darwinian scientific ideas were repugnant to religious authorities—auguring the brewing conflict between science and faith that would intensify and continue to our day. And there were other problems with the classification system. Linnaeus's lowest level for humans was race; he defined the human races as Africanus, Americanus, Asiaticus, and Europeanus. This classification gave rise to the idea of race-related distinctions among people.

Despite the conflict between entrenched belief and his new way of seeing the world, Linnaeus thrived. In 1741 he was named chair of medicine at Uppsala, and he later took over the chair of botany at the university. In 1761, King Adolf Fredrik ennobled him in recognition of his important contributions to science. Linnaeus continued his work with great passion—he knew that he was developing a new science and that he was extending comprehension of the living world through the connections and distinctions he was uncovering among living creatures.

These ideas of classification, which were sweeping the intellectual world in the wake of Linnaeus's discovery, would inspire Charles Darwin in the next century and lay the groundwork for a new way of thinking, at odds with the traditional belief in a static world divided into humans on one side and all other creatures on the other. Associated with this biblically inspired belief was the social understanding that the European economic classes were intrinsically different from one another. The rich were rich because God and nature deemed for them to be so, while the poor belonged to a low class, and were forever to remain there; princes and kings, and paupers as well, were ordained by God. Linnaeus's simple classification system of all living things reflected the downfall of this medieval way of thinking, as culture and the economy were evolving.

Of course, the people who benefited most from the prevailing false social classifications felt threatened. But history was to move forward, and this was the Age of Enlightenment, when science and reason were prevailing over age-old beliefs.

Linnaeus's system never said anything about evolution. It was a system of classifying living things; it did not imply that one species evolved from another, living or extinct. In fact, Linnaeus believed in the fixed nature of his system. Observing connections among species did not lead him to assert a way for these connections to have been forged through mutations over time. His work did imply, however, that the natural world had an inherent *order*, and that this order could be deduced by using the scientific method. This method could employ anatomical and botanical analyses of specimens to uncover their properties and thus deduce the rules that produced them: the laws of nature. The mere idea that the laws of the universe could be understood, in the sense that people could see how organisms were related to one another, was crucial for developing the theory of evolution.

The hobby of collecting butterflies originated at this time, and Europeans went to the seashore to search tidal pools for interesting life forms. Museums of natural history were inaugurated, and natural history societies were founded in many cities. Occasionally amateurs would also find fossils, but generally they would not make the connection between extinct forms of life and what they were finding. A fossil shell was assumed to be an unusual piece of rock that resembled a shelled organism. A fossilized branch was not understood to have once been part of an ancient plant or tree—it was just a curiosity.

The assumed static character of the natural world, as dictated by the biblical story of creation, was still unassailable. According to Genesis, the world was created with all its creatures—there was no allowance for a dynamic element with species evolving and dying out. Everything living in the world had always been there, and would always remain.

Had people believed otherwise, not only would they have contradicted the biblical account of creation, but such an idea would have constituted a challenge to divine wisdom. For why would the creator need to "experiment" with life? His perfection implied the perfection of all creation and allowed for neither evolution nor extinction as time progressed. The world had a long way to go to understand that biblical notions should be taken only allegorically. It had required centuries for religious people to accept the Copernican theory, Kepler's discoveries, Galileo's work, and finally Foucault's definitive proof of Earth's rotation.

So while the Western world mostly welcomed Linnaeus's idea of classifying all living things and embraced the scientific revolution enabled by his research, there was a limit to how far interpretations of his work could go. But the accumulating fossils being discovered by amateurs and scientists were pressing to tell their story.

More than any other country, France was striving to establish itself as the world leader in science, especially natural science. Paris's Museum of Natural History was founded on June 10, 1793, at the time of the French Revolution, on the site where, in 1635, King Louis XIII's physician had created a royal medicinal garden (now the Jardin des Plantes). The French government, in all its post-revolution phases, continued to fund this institution generously and established a number of endowed chairs in an effort to promote the study of nature. By the nineteenth century the museum came to rival the University of Paris in prestige as a research institution in natural science.

One of the newly established professorships at the museum was held by a master anatomist named Georges Cuvier (1769–1832). In 1796, Cuvier wrote a paper about living and fossil elephants. He analyzed the skeletons of living elephants from Africa and Asia and compared them with fossils of mastodons, extinct elephants that inhabited North America until about 13,000 years ago. His paper spelled out the startling view that fossils were the remnants of living things. He thus also introduced the idea of the extinction of species: Creatures lived for generations, and at some point, for reasons he could not explain, they ceased to exist. Cuvier understood that the world contained both living creatures and the fossilized remains of species that had disappeared forever.

Since religious leaders and their followers refused to believe that extinction was possible—the idea, after all, contradicted the perfection of God—an explanation for this phenomenon was needed. It was proposed that some fossils represented animals that no longer existed at the location at which they were found, but continued to exist as living organisms elsewhere on the earth. Explorations during this age of war and political expansion could perhaps solve the mystery of whether animals were indeed extinct in one location but alive in another, and explorers and adventurers set off in search of extinct creatures that might be alive in other parts of the world: mammoths in

North America, shelled organisms on faraway ocean shores. But none was found.

Cuvier's ideas were at odds with those of a fellow professor at the Museum of Natural History, Jean-Baptiste de Lamarck (1744–1829). Lamarck had studied mollusks, and his analysis made him theorize that living things transmuted over time: Simple organisms changed gradually into more complex living things. Lamarck did not believe in extinction; species, he said, changed into more complex ones. While he was wrong about extinction, Lamarck was very perceptive about evolution—which he called transmutation—although he had no understanding of how the process actually worked. He hit upon the concept of evolution long before Darwin, but the exact mechanism that made it work eluded him. He believed, for example, that if generations of mice had their tails cut off, eventually mice would be born tailless. According to this theory, Jewish men and others whose ancestors have been circumcised over many centuries would by now be born circumcised. But Lamarck did have the basic idea of evolution. The explanation of the process would be the work of Darwin. Lamarck also revised the Linnaean classification system, developing it further to lower-level animals—an area that Linnaeus had not attended to.

Lamarck exploited the medieval idea of the *scala naturae*, "nature's ladder," which claimed that all living things can be arranged on a scale of increasing complexity from most primitive to most perfect: up to people and on to angels. This system had originally been used to support the idea of immutability of species—the fixed position of each living creature within the universe. Lamarck adapted it to his own purposes, proposing that within this system, every creature strove to improve itself, and thus to climb the *scala naturae*.

Cuvier, in contrast, did not believe in transmutation. He thought that Lamarck's ideas were silly and not based on any empirical evidence. He believed that species became extinct through a series of catastrophes such as the biblical flood. He did not question the static biblical assumption that all creatures stay in their own places, and in order to explain the extinctions he saw, he proposed a system by which the flood played a repeating role through history. Cuvier and Lamarck attacked each other in print, and no one could decide which of the two was right. In a sense, both were right and both were wrong.

Another important development in preparing the world for the concept of evolution came in England toward the end of the eighteenth century, when William Smith (1769–1839) produced a map that showed the structure of the surface of the earth. Smith studied the geology of Britain and discovered various strata and the fossils that appeared within them. He coined the term "faunal succession" to describe the sequence of fossils he found at varying strata from deep in the earth up toward the surface. From his geological work digging in the ground, he understood that the layers of sedimentary rock represented time sequences, with older layers on the bottom and younger ones above. Smith concluded that since one could dig deeper and deeper, and thus move further back in time, the planet must be much older than biblical scholars had projected. Equally, since more primitive forms of fossil life were discovered at greater depths, it was clear to Smith that life had progressed in complexity through time.

In Smith's day and before, the biblical story of creation led some people to believe that the world was less than 6,000 years old. By counting the "begats" in the Bible all the way from Adam to Jesus, and making some other calculations, James Ussher, the archbishop of Armagh, in Ireland, deduced in 1650 that creation took place at nightfall preceding Sunday, October 23, 4004 B.C., by the Julian calendar. (Some sources give the time as nine a.m., and some attribute the determination of the moment of creation to the churchman John Lightfoot of Cambridge University.) Ussher's calculation is still believed by some people to be correct.

Even without the benefit of modern dating techniques, which are based on the known rate of decay of radioactive elements such as uranium (used for long spans into the past) or the isotope carbon 14 (for up to 50,000 years with reasonable accuracy), Smith understood that the rocks he was studying, found at deep locations under the surface of the earth, were many orders of magnitude older than 6,000 years.

Later, this discovery of the great antiquity of rocks and their increasing age as one moved deeper into the earth would help scientists analyze and estimate the age of fossils found within various geological strata. Just as important, Smith's discovery of the correlation between the complexity of fossilized life forms and the time sequence of the geological strata where these fossils were found helped pave the way for the theory of evolution. The discoveries made and the theories proposed by Linnaeus, Cuvier, Lamarck, and Smith by the end of the eighteenth century were harbingers of the revolution whose time had by then arrived. This revolution would be the prize of Charles Darwin.

Chapter 3

DARWIN'S BREAKTHROUGH

When he read Darwin's *On the Origin of Species*, the naturalist Thomas Henry Huxley (1825–1895), a scientific visionary on a par with Darwin, exclaimed: "How extremely stupid of me not to have thought of that!" Eventually, it was Huxley, who understood the depth and implications of Darwin's theory better than did most of his contemporaries, who helped bring about the widespread public acceptance of Darwin's ideas of evolution, adaptation, and natural selection. But when Darwin's book was first published in 1859, evolution seemed a ridiculous idea to most people. The main problem with the theory of evolution was that it contradicted the literal interpretation of scripture, which states that man was made on the sixth day of creation, after the earth, sky, water, trees, plants, and animals had all come into being by divine decree. Science had been progressing, with new discoveries about nature being made every year, and by the time Darwin was ready to publish his theory—which was more than two decades after he had developed his ideas—evolution, as a concept, was almost "in the air."

Charles Robert Darwin was born on February 12, 1809, in his family home, called Mount House, in Shrewsbury, Shropshire, England. He was the fifth of six children of a wealthy doctor, Robert Darwin, and his wife, Susannah Wedgwood, of the famed Wedgwood pottery-making family. His paternal grandfather, Erasmus Darwin, had proposed a theory similar to Lamarck's about biological evolution, in *Zoonomia; or, The Laws of Organic Life*, which though not very scientific, and replete with speculation, contained a germ of the idea of evolution. It did not generate much public interest. Charles's mother died when he was eight, and the boy was sent to boarding school. When he grew up, he wanted to be a doctor like his father, and enrolled to study medicine at the University of Edinburgh.

Darwin didn't much like the study of medicine, and he found surgery too bloody. He joined the university's naturalist society, where he heard about travel to exotic lands. Through his professor of natural science, Robert Edmund Grant, Darwin learned of Lamarck's ideas on evolution, as well as those of his own grandfather Erasmus. He took part in Grant's research on marine mammals and also analyzed oysters. In addition, he studied the new concepts that were gaining favor in geology, and learned about stratigraphic analysis and the dating of geological formations.

In 1827, Darwin transferred to Christ's College, Cambridge, to study theology. But here he found himself spending most of his time collecting beetles, riding, and hunting. He did very well in natural history, geology, and botany—and in theology. After graduation, he was recommended by his botany professor for an unpaid position of gentleman's companion to Robert FitzRoy, the captain of HMS *Beagle*. This ship was about to embark on a two-year expedition to chart the coastline of South America, and his professor thought it would be an exceptional opportunity for Darwin to learn firsthand about natural history. As it turned out, the *Beagle* traveled for five years around the world, rather than two around South America, and the journey would change forever how people viewed the natural world.

Throughout this long voyage, Darwin spent much time ashore at every port of call, exploring uncharted shorelines, digging for fossils, and collecting an enormous number of specimens of animals and plants, a large proportion of them unknown to scientists in Europe.

Darwin took with him on his voyage a copy of the seminal *Principles of Geology* by Charles Lyell (1797–1875). The book inspired him with its explanation of stratification into geological layers and of how these levels allowed an estimation of age, which in turn could explain the kinds of fossils in them and lead to the observation that the fossils of more complex organisms were generally found at the younger, upper geological layers.

Lyell's book also explained the action of the forces of nature—rain and flooding, wind erosion, volcanic activity—and their impact on the geological landscape over very long time spans. These forces could bring lower layers up to the surface, thus changing the terrain. This was useful information for the young naturalist, as it aided him in identifying various finds and explaining the natural laws ruling our planet.

In Patagonia, in southern Argentina, Darwin observed a stepped plain with seashells, and with Lyell's explanations, he understood that what he was seeing was actually a beach that had been raised above sea level by geological forces over millions of years. In Chile, he found mussel beds that had been similarly stranded on a hill by an earthquake. He found fossils of giant extinct mammals, and in Argentina discovered two species of rhea, a South American flightless bird resembling an ostrich but smaller, that had separate yet overlapping territories. In Australia he marveled at the beauty of marsupials.

But it was the Galápagos Islands that offered Darwin his best natural laboratory, and what he saw there precipitated his most profound thoughts about evolution. He found that mockingbirds on one island were different from those on another, and thus arose the idea that separate paths might be taken in the process of evolution. He saw that tortoises and finches also differed among their separate islands, reflecting the same process of evolution acting along different lines in separate environments.

During his long voyage, Darwin was often sick with fever, stomach pains, palpitations, and tremors; for one period he had to spend an entire month in bed. These symptoms plagued him throughout the rest of his life, and at times would be attributed to stress. It has been suggested, too, that he caught a tropical disease, which kept recurring later in life, induced perhaps by stress or other causes.

When the *Beagle* returned to Britain on October 2, 1836, Darwin's father and his professors at Cambridge promoted his work. He was hailed as a naturalist who had traveled to many unknown places in the name of science, and as a result of this publicity he was invited to lecture and make presentations of his findings. Darwin persuaded a number of his former professors at Cambridge, and faculty at other universities, to study the many specimens he had brought with him from his voyage. An analysis of the fossils at the Royal College of Surgeons in London revealed that some of the bones he had brought back belonged to extinct giant rodents and sloths. This caused great excitement in the scientific community, as did the large collections of stuffed mammals and birds that Darwin donated to the Zoological Society of London.

In 1837, Lyell devoted his presidential address before the Geological Society of London to discussing Darwin's results—especially the fact that he had found fossils of animals in the same locations where similar living animals now roamed the land, which led him to theorize that these creatures had evolved over time. Darwin continued to lecture; privately he wrote notebooks in which he explored the ideas he was not ready to disseminate to the public. They included his theory that the various tortoises of the Galápagos had all originated from a single species that later adapted to different environments on different islands, with variations that were due to environmental adaptation.

For his achievements, Darwin was elected secretary of the Geological Society in 1838. Eventually he published the journal he had kept during the voyage of the *Beagle*. Before then, though, he decided to get married. He proposed to his cousin, Emma Wedgwood, and when he told her in secret of his ideas about the transmutation of living things—the beginnings of his theory of evolution—she became worried. Emma thought that if they married, they might not be allowed to meet in paradise once they both were dead, since obviously this young man's ideas were not in line with Christian belief. To allay her fears, Darwin wrote Emma a mollifying letter. She accepted his reassurances, and in January 1839 they married. Days earlier, Darwin had been elected a fellow of the Royal Society, the world's oldest scientific academy in continuous existence.

Since both their families were wealthy, Darwin and Emma did not have to worry about his livelihood. They later settled in Down House, Kent, not far from London, and raised a large family: ten children, of whom three died young; others suffered from illnesses. Darwin himself was still plagued by bouts of poor health, which he tried to cure with long periods of rest in the countryside.

As a celebrated and highly honored scientist, Darwin was reluctant to jeopardize his standing with any controversy—so he kept working on his theory of evolution in private. He knew that if his ideas were published, a great debate would arise. Anglican religious authorities, whom he knew well from Cambridge, would not take his theory well and would try to discredit it and him. He had only to lose from the dissemination of his dangerous notions about the evolution of living creatures. And he knew that other elements in British society would object to his ideas. People who were comfortable with their place in society would feel uncomfortable in a system that might now be seen as inherently in flux.

DARWIN'S IDEAS were influenced in part by the brilliant Scottish geologist Roderick Murchison (1792–1871), who believed that there were several ages of the earth. First came an age without any organisms. There followed the age of invertebrates, then the age of fishes, and then that of reptiles; next came the mammals; and finally there was man. This theory of ages brought about a general notion of geological eras. Darwin now believed, on the basis of Lyell's and Murchison's works, that the planet was very ancient. It had undergone these various ages, going back eons, and life had evolved through them.

Darwin's notebooks make it clear that he was coming to believe that ordinary events took a very long time to manifest themselves. He thought that, as in geology, with its very slow processes of erosion, almost imperceptibly slow changes were taking place within living organisms. These changes in living creatures made themselves evident only after very long periods. Geology thus provided Darwin with a model for biological change. Rocks and animals, in a sense, were undergoing analogous transformation through time.

The mechanisms for producing this change-through-time in living things were natural selection and survival of the fittest. Organisms evolved: Members of a species that were best adapted to their environment survived in greater numbers and were thus naturally selected to breed and produce progeny more frequently and more effectively than organisms that were less well adapted to their environment. As time passed, the better-adapted organisms survived and dominated.

Darwin used the expression "transmutation of species" for this process, which later became known as "evolution." He began his analysis by collecting the many facts he had deduced from observations made over his five-year journey with the *Beagle*. He included the results of studies of his data that had been conducted by botanists, zoologists, and geologists. He spoke with breeders of horses, dogs, and other animals and learned about the changes that occur in individuals as generations of animals are born, breed, and die. Darwin was building his case for evolution; he was careful with every detail, to avoid the possibility of ridicule or dismissal, as Lamarck and his own grandfather Erasmus had endured when attempting to construct a system of evolution without the required data to support their theories.

As Darwin recalled much later (in 1877), when he was on the *Beagle* he still believed in the "permanence of species," but doubts crept into his mind when he made new observations. On his return home in the autumn of 1836, however, he wrote, "I immediately began to prepare my journal for publication, and then saw how many facts indicated the common descent of species, so that in July 1837 I opened a notebook to record any facts which might bear on the question. But I did not become convinced that species were mutable until, I think, two or three years had elapsed."

In 1838, after reading Thomas Malthus's essay on populations, Darwin began to think more acutely about how populations grow, and this helped precipitate in his mind the key principle of evolution: natural selection. By 1842 he had written his ideas and would have started to consider publishing them. But Darwin, still concerned with how his theory might be received, shied away from publication, and the project was stalled.

Darwin had been unaware that he had a competitor, Alfred Russel Wallace (1823–1913), an English naturalist who in 1855 published a paper titled "On the Law Which Has Regulated the Introduction of New Species," which set down the ideas of an evolutionary theory similar to Darwin's. In early 1858, Wallace even sent Darwin a letter with his ideas about evolution by natural selection from the Moluccas, in the Malay Archipelago, where he was bedridden with malaria. When Darwin learned that someone else was about to be credited with the discovery of a theory of evolution, he finally decided to publish his ideas. His book on evolution, *On the Origin of Species by Means of Natural Selection, or The Preservation of Favoured Races in the Struggle for Life*, appeared in print on November 24, 1859. Darwin had in his possession the notebooks he had begun to write two decades earlier, and he was able to use them to prove that he was the first to propose a theory of evolution. But already in 1858, at the urging of friends after he received Wallace's paper, Darwin had made a presentation of his theory together with Wallace, who presented his own findings, at a meeting of the Linnean Society in London.

There are several key elements to Darwin's theory of evolution as presented in *The Origin of Species*. First is the element of time and change. The living world is a changing world: Species change over time, and new species evolve while others become extinct. According to Darwin, this process is gradual and continuous. Today there are different interpretations of exactly how species evolve (the process may be faster than Darwin imagined).

Another key element of Darwin's theory is common descent: All mammals have a common ancestor, and similarly, there is an ancestor to all birds, and to all reptiles, and so on. The totality of life on the planet, both plant and animal, may have a common ancestor as well.

Further, the process of evolution is driven by natural selection: Those individuals in a population that have the best characteristics in the sense that they are best adapted to their environment—they can find food and shelter and mates, and avoid predators, better than others—will tend to reproduce more often, and their useful characteristics will pass on to their descendants. Everything is transient in this life, generation giving way to generation, and offspring are modified through the powerful mechanism of natural selection. The present is a museum of the past, but it is an imperfect museum—it contains only what we can see; it is "a poor collection made at hazard and at rare intervals." According to Darwin, we must abandon the idea of a permanent identity, and "regard every production of nature as one which has had a history." Applying Darwin's ideas to human beings, we see that we, too, are subject to the vicissitudes of time, and can be understood as a species only when we view ourselves within our relationship to the past.

Yet another important element of Darwin's theory is the notion that every population has a natural variation in the characteristics of its members. Natural selection works by favoring those individuals within this statistical variation that are best suited, best adapted, to the environment in which they live at a given time. Once the environment is altered, either by climatic change or by migration into new territory, again those individuals that are best adapted to their new conditions are the ones that survive and breed and thus create statistically better-suited offspring.

Darwin's *Origin of Species* contained ideas that went beyond the survival of a single individual. Darwin wrote that he used the term "struggle for existence" in a "large and metaphorical sense, including dependence of one being on another, and including (which is more important) not only the life of the individual, but success in leaving progeny."

The idea that evolution may be extended from an individual to a society, and perhaps to a species or several species together—and even to the entire planet, viewed as a single living mega-organism—would be taken up in the twentieth century. The person who would do this was Pierre Teilhard de Chardin, extending Darwin's idea of evolution to our planet as a whole, and taking the concept beyond, to a higher spiritual realm. And like Darwin, he would be strongly criticized and would pay a heavy price for his ideas.

When it appeared in 1859, *On the Origin of Species* at once reinforced speculation on human development and, more important, provided a theoretical framework within which this process could be explained. As fossils of what seemed like forms of humans—although quite different in their skull shape and bone structure—were later being discovered in Germany, France, and Spain, science had the theoretical framework to interpret them.

The controversy that Darwin feared for so many years did indeed erupt—and it continues to this day. On June 30, 1860, seven months after the appearance of Darwin's book, the "Great Debate" was held at

Oxford University. This major face-off about human evolution pitted the bishop of Oxford, Samuel Wilberforce, against Thomas Henry Huxley, then a professor at the Royal School of Mines in London. One observer described the latter: "A slight tall figure, stern and pale, very quiet and very grave, he stood before us and spoke those tremendous words—words which no one seems sure of now, nor, I think, could remember just after they were spoken, for their meaning took away our breath."

Like many in the Anglican Church, Bishop Wilberforce was horrified by the suggestion that humans did not come into existence by God's act of creation of Adam and Eve, but were descended from apes or their ancestors. Before his meeting with Huxley, who was such an ardent advocate of evolution that he called himself "Darwin's Bulldog," Bishop Wilberforce had been coached by Richard Owen. Owen, an eminent scientist, was a strict believer in creation, but was familiar with evolutionists' arguments. Owen and Huxley had been, and would remain, the main antagonists in the controversy between evolution and creation.

Wilberforce opened the debate with a long speech about evolution and creation, at the end of which he turned to Huxley and asked a question intended to elicit ridicule: If we are all the descendants of apes, then on which side, his father's or his mother's, was Huxley descended from an ape?

Upon hearing this question, Huxley reportedly turned to his neighbor and whispered: "The Lord hath delivered him into my hands." He then rose to his feet and responded to Wilberforce. He said that he would not be ashamed of his ancestry. He would, he said, only "be ashamed to be connected with a man who used great gifts to obscure the truth." As was clear to all present, Huxley was saying, in fact, that he would rather be related to an ape than to a misguided bishop. The audience was stunned by his audacity, and pandemonium ensued. A witness reported that "no one doubted his meaning, and the effect was tremendous. One lady fainted and had to be carried out; I, for one, jumped out of my seat."

Over the years and decades, this debate would widen and intensify. The discoveries of fossil hominids, which began roughly around the time of Darwin and have continued with increasing frequency into our own century, add important substance to this debate, as do new developments in genetics and molecular biology.

Along with Darwinism, there came the concept of a "missing link" between humans and apes—even though in *The Origin of Species* Darwin was careful not to deal directly with human evolution, and did so only in *The Descent of Man, and Selection in Relation to Sex*, published in 1871.

In 1877, Darwin was awarded an honorary degree from Cambridge University. The students prepared a prank, in which, just as the conferring ceremony was about to begin, a monkey marionette was brought out on a cord that had been strung above the audience. This caused a great uproar. The monkey was followed by a ring with ribbons tied all around it. This ring was meant to represent the "missing link."

Today we understand that evolution works its magic not within large populations—which tend to be stable over time—but rather through the event of speciation, the formation of new species. This process usually acts on a small subpopulation that is separated from the larger population of which it was once part, away from the domain of the parent population. In a new environment, isolated from its parent, the smaller group is subject to the evolutionary forces of adaptation and natural selection, and these forces bring about genetic change in a relatively short time—thousands of years, a mere instant in geological terms. Those individuals in the isolated population that adapt best to the new environment reproduce more effectively and yield offspring that favor these adaptations. It was through this process that a group of hominids belonging to the species we call *Homo heidelbergensis* (discovered both in Africa and in Europe, near the German city of Heidelberg, and dated to several hundred thousand years ago) left Africa, their ancestral land, and found themselves on the much colder European continent—a land dominated during the Ice Ages by glaciers, snow, and permafrost. The forces of evolution, acting on this population hundreds of thousands of years ago, produced a new hominid species that was uniquely adapted to cold weather: the Neanderthals, or *Homo neanderthalensis*.

Chapter 4

STONE TOOLS AND CAVE ART

Throughout history, people have discovered fossils and curious stones that evidently had been worked on: broken off from a larger rock and then systematically chipped all around to give them sharp, solid edges. Such worked stones actually constituted the various prehistoric tools used by humans and other hominids to hunt, skin, and butcher animals, and they have been found in very large numbers. These fossils and stone tools have been discovered in digs and in caves, and sometimes just strewn on the ground—since the forces of nature may bring to the surface ancient layers from below.

The first recorded discovery to indicate the antiquity of man was made only in 1790. That year, an English farmer named John Frere found sharp stone tools, along with fossils of extinct animals, in a gravel pit near Hoxne, in Suffolk. Frere understood that these animals were very old and that the tools implied that humans were contemporary with the ancient animals that once lived in the area. He subsequently published a paper describing the stone tools, saying that the people who made them "had not the use of metals." But his discovery and his paper were largely ignored.

Similar finds that should have indicated that humans and other hominids had lived on Earth a very long time ago—because their remains or their stone tools were located among fossils of extinct animals—were made in the early nineteenth century. But none was recognized as such. The fossils discovered were of anatomically modern prehistoric humans, who lived more than 10,000 years ago, and of Neanderthals—members of what is known today to be a separate and distinct early human species, now extinct. The Neanderthal fossils included cranial fragments discovered in 1829–1830 at the Engis cave in Belgium, now known to have belonged to a Neanderthal child of about three; and a strange-looking skull discovered in 1848 at the Forbes Quarry, at the site of a prehistoric rock shelter on Gibraltar, later taken to the Gibraltar Museum, and known today to have belonged to a female adult Neanderthal.

IN AUGUST 1856, the great discovery was made—one that signaled the birth of paleoanthropology and led to our understanding that the globe was once inhabited by members of the human family that were not quite like us. This monumental find at a limestone quarry overlooking the Düssel River in Germany's green, pastoral Neander Valley (Neandertal or Neanderthal in German) provided the theory of human evolution with its first solid piece of evidence.

Because it was hard to reach, one cave in the quarry, situated on a steep ridge above the river, remained unexploited. But finally its time had come. Workers climbed up the ridge with ropes, entered the narrow cave, and laid explosives in place, and the foreman gave the order for the blast.

Once the dust from the explosion had settled, workers set to clearing the rubble. One of their spades hit an unexpectedly hard surface, and on inspection they saw a skull. Not far behind it were found a pelvic bone and thighbones. The foreman assumed that these were the remains of a bear. He happened to know a high school teacher in the area, Johann Carl Fuhlrott, who was an amateur naturalist and had studied natural science at the University of Bonn. Sometime later, when the work was finished, the foreman contacted Dr. Fuhlrott and invited him to the quarry to collect some bear bones that had been put aside for him. Fuhlrott came to pick up his prize.

Fuhlrott knew enough anatomy to understand that the fossilized bones in his hands did not belong to a bear. They looked human—but not quite. The skull was more elongated, and flatter at the top than a modern human skull. Fuhlrott excitedly concluded that the remains were those of a human ancestor—a creature distinct from a modern human yet similar to it; it must have been an earlier human species, now extinct.

At first, no one believed Fuhlrott's conclusions. Anatomists argued about the strange-looking fossils. A German professor thought they were the remains of a Cossack who had fought in the Napoleonic Wars, was wounded, crawled up into the cave to seek shelter, and died. A French scientist thought the remains were those of a Celt with a deformed skull. Another expert thought the bones were those of a retarded modern individual. The Neanderthal find was about to be forgotten.

Fuhlrott, eager to confirm his own interpretation, consulted Dr. Hermann Schaffhausen, a professor of anatomy at the University of Bonn. Schaffhausen concurred with Fuhlrott's assessment that the fossils represented an ancient form of human, and agreed to make a joint statement about the find and its significance. After their announcement, Fuhlrott and Schaffhausen were criticized; their interpretation was considered contrary to the scriptural doctrine of human creation.

Today, as noted above, we know that the 1848 Gibraltar find belonged to a female Neanderthal, and that the child's skull from Engis discovered in 1829–1830 was also Neanderthal. This species of human lived between about 300,000 and 30,000 years ago—when it mysteriously disappeared. These hominids lived mostly in Europe, but also, during limited periods of time, in parts of western Asia.

In 1864, the Irish anatomist William King took a major step forward

in science and named the species to which all these finds belonged
Homo neanderthalensis. This binomial meant that Neanderthal was a
hominid—a member of our own genus, *Homo*—yet constituted a distinct species within the genus.

Further discoveries followed, and they were identified as belonging
to this newly defined member of the human family. These discoveries
included another in Belgium, at Spy, in 1886; finds in Krapina, Croatia, between 1899 and 1906; and another in Germany, this time near
Ehringsdorf, in 1908. There were many discoveries in France between
1908 and 1914, and in the Crimea in 1924–1926. Then there were
important Neanderthal finds in Israel, in the caves of Mount Carmel
and the western Galilee, starting in 1929, when the noted British prehistorian Dorothy Garrod (1892–1968)—the first woman to hold a
chaired position at Cambridge—and her team excavated the cave of
Tabun. Garrod's discoveries saved Tabun and other caves from destruc-
tion, for the British Mandate government wanted to turn the cliffs of
Wadi el-Mughara on Mount Carmel—since then found to be rich in
prehistoric fossils—into a giant quarry for building the harbor at
Haifa. There were also finds in Italy and Uzbekistan in the 1930s, and
in the 1950s at Shanidar in Iraq.

One of the fossil skeletons found at Shanidar was especially interesting. With the skeleton, which seemed to have been buried, scientists
discovered extensive remains of flower pollen. This led them to
hypothesize that flowers had been placed with the body when it was
buried, and that Neanderthals had something akin to religious belief,
or at least a symbolic remembrance of the dead.

Neanderthals have not been found in Africa or eastern Asia. These
hominids were specifically adapted to the glacial climates of Ice Age
Europe, with thick bodies, which preserved heat much better than
those of modern humans, and large, bulbous, protruding noses whose
purpose was to help warm the freezing air before it entered their
lungs. They had stout limbs, and their bone fossils show signs of heavy

musculature. Their lives entailed extreme exertions—their shinbones have been shown to withstand a stress level three times as high as the level that modern shinbones can take. Their knees and ankles were larger and much stronger than ours.

These peoples lived and hunted in the icy conditions of Europe and western Asia, using stone tools they produced to hunt and butcher animals. Some scientists have hypothesized that the Neanderthals came south to Israel and Iraq only when the glacial climates of Europe became too cold even for their cold-adapted bodies, and inhabited the warmer Middle East only temporarily.

So far, finds belonging to more than five hundred individuals have been discovered over the entire range of these hominids. These discoveries comprise fossil bones and parts of skeletons, skulls and parts of skulls, and sometimes almost complete skeletons. Along with these remains, hundreds of thousands of stone tools have been found. The Neanderthal's brain was larger, on average, than our own. The average cranial capacity of a modern human is about 1,400 cc (cubic centimeters); the Neanderthal brain averaged 1,500 cc. But how did their intelligence compare with that of our anatomically modern human ancestors? And what was the relationship between these two members of the human family?

IN 1863, along the Vézère River in the Périgord region of France, below a medieval rock shelter called La Madeleine, excavators found a large number of stone tools, along with human and animal fossils. This unexpected discovery of prehistoric remains, which was made while archaeologists were excavating an eighth-century rock dwelling, led scientists to define a new concept: the Magdalenian culture, a Stone Age society named after this location, which was believed to have lasted from about 18,000 to 11,000 years ago. The people who made these tools were not Neanderthals, but anatomically modern humans. The prehistoric Europeans are now called Cro-Magnons, after the discovery of the remains of such people in 1868 in the cave of Cro-Magnon, near the village of Les Eyzies-de-Tayac, also in the Vézère River Valley, not far from La Madeleine.

The Cro-Magnon people lived from about 40,000 to about 10,000 years ago; their prehistoric communities eventually gave way to early agricultural societies. Unlike the Neanderthals, which form their own species, *Homo neanderthalensis*, the Cro-Magnons belong to our species, *Homo sapiens*. Anatomically identical to us, the Cro-Magnons possessed a chin (which Neanderthals did not; their lower jaw lacked the forward protrusion that forms the chin), and therefore, at least theoretically, had the potential for speech. Did they have a language, or several languages? And did they possess the ability to think symbolically? If they had languages, none of these is known to have survived as such (although some scientists believe that the Basque language, which is not Indo-European, may have Ice Age roots). But we do have some marvelous evidence of their symbolic thinking.

In 1875, a shepherd told the Spanish naturalist and amateur archaeologist Marcelino Sanz de Sautuola about the existence of a cave near Altamira ("high lookout"), a few miles from the Atlantic coast near Santander in northern Spain. Sanz de Sautuola was interested in prehistory, and he visited the cave and saw a few artifacts, but nothing noteworthy. Three years later, he went to the Universal Exposition in Paris, and it fired his imagination, for there he saw prehistoric objects from the site of La Madeleine, which were similar to some that he had seen in the cave of Altamira. He returned to explore Altamira, and it was then that he discovered the extensive cave art: images of bison, horses, and deer, painted in yellow, red, and black on the cave ceiling.

Many other caves with Stone Age art have since been discovered—a large number of them in France, in the Périgord as well as in the Pyrenees and the Ardèche. These include the Lascaux cave, found in 1940, and the Chauvet cave, found in 1994. Both have substantial and extremely important art, yielding hints about how the Cro-Magnons thought—although much that is behind this art remains a mystery. Among breathtaking red and black pictures of bison, horses, elk, and reindeer, there are also recurring symbols whose meanings have not been deciphered. In France alone, there are more than two hundred known caves decorated in some way: by paintings on the walls or ceiling, wall carvings, or statuettes found within.

As a young paleontologist, Pierre Teilhard de Chardin visited many caves, and he studied the artifacts found there. His close friend the Abbé Henri Breuil (1877–1961), a Catholic priest who was also a professor of prehistory, made some of the most important contributions to the study of cave art, including authentication of the antiquity of the paintings and drawings found at Lascaux. Starting such work in 1901, when he co-discovered the wall paintings in the cave of Font-de-Gaume in the Périgord and in the cave of Mas d'Azil in the Pyrenees, Breuil continued his study of prehistoric art and, as an expert in this area, authenticated the paintings and artifacts found in most of the significant caves in France. He hand-copied and published most of the drawings and paintings found in the caves of France and Spain during his lifetime.

The anatomically modern humans who produced the cave art of Europe were doubtless able to think symbolically. The cave art testifies to this. The subjects of this art are almost always animals, and the drawings are of specific styles, which recur over millennia and across the terrain. The most common animal is the bison, which was abundant in Ice Age Europe, often accompanied by this sign:

Other animals commonly depicted are horses, reindeer, and mammoths. Lions, rhinoceros, cave bears, fish, wolves, ibex, and other

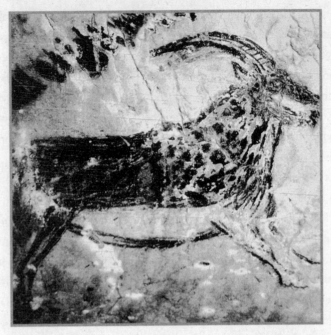

Drawing of an ibex, Magdalenian period (about 15,000 years ago), from the Niaux cave, France. Conseil Général Ariège Pyrénées. Photograph by Debra G. Aczel.

animals occur less frequently. Birds appear very rarely, and there is never any recognizable kind of scenery whatsoever. There are some vague "anthropomorphic" figures, but no detailed images of human beings—while the animals are painted or drawn, in charcoal, in minute detail. But each animal is shown without any apparent connection to terrain. Often, the images of these animals are superimposed on each other; for example, a bison might be superimposed on a horse, sharing a tail or some other feature.

Symbols accompany most cave art. In the Gargas caves of the French Pyrenees are more than 150 paintings of human hands in various positions, radiocarbon-dated to 27,000 years ago, which are tantalizingly suggestive of a language. Other prehistoric symbols include rows of dots, which appear in many caves, often close to the entrance:

And another recurring symbol is a pair of lines:

No one knows the meaning of the symbols, or the purpose of the extensive cave art with its very specific nature. Did the animals and symbols serve a religious purpose? A form of communication? Art for its own sake? Did they delimit tribal territory? In 2001, a German researcher claimed that he had solved the mystery: The dots at Lascaux represented the phases of the moon; he counted twenty-nine of them and thus hypothesized that the pattern represented the twenty-nine-day lunar cycle. But there are many dot patterns at Lascaux alone, and it is not clear which dots should be counted. And there are many other dot patterns in other prehistoric caves where the number of dots is not twenty-nine. So the meaning of the magnificent prehistoric cave art remains a mystery.

Usually, caves with extensive art have not been found to contain other signs of human habitation: no fossils, no stone tools, no animal remains, no signs of cooking or eating or butchering animals. The only signs that people were there, other than the art itself, are carbon remains from the burning of stone candles fueled by animal fat, which the Cro-Magnons apparently used for lighting the caves while creating their art. Caves with a great deal of art in them were never used for living. And the paintings and drawings were often hidden in the most remote and inaccessible parts of the caves. In the case of the Niaux cave in the French Pyrenees, the first gallery of drawings, called the "Black Salon," is about half a mile (800 meters) deep inside the cavern. For this reason, while graffiti in this cave indicate that parts of the cave close to the entrance had been discovered as early as 1602, the drawings were not seen until three centuries later, when, on September 21, 1906, the brothers Paul and Jules Molard and their father found the marvelous drawings of bison and horses and an ibex; they later had the Abbé Breuil authenticate the drawings as prehistoric. The artists aimed—and succeeded—at hiding their drawings well.

While nobody knows for sure, it is believed—because of the proximity of caves with art to caves where fossils were found—that all cave art was done exclusively by the Cro-Magnons; none was made by the Neanderthals. Stone tools, however, were produced by both species, and they are almost identical; their specific type depends on the Stone Age culture to which they are seen to belong. The Mousterian toolmaking industry is named after the French village of Le Moustier, three miles northeast of La Madeleine in the Vézère River Valley, near where these tools were found in a rock shelter. This tool-making culture lasted from 250,000 to 40,000 years ago.

Before the Mousterian industry, and after it, came other stone industries, whose tools are characterized by their shape and size; the tools are found in Africa and Europe, as well as Asia, and include hand axes, scrapers, and points. Among these stone industries are the Magdalenian, 18,000 to 11,000 years ago; the Acheulean, 1.5 million to about 200,000 years ago; the Oldowan (named for the Olduvai Gorge), the oldest industry, starting as far back as 2.5 million years ago in Africa. Between the Mousterian and the Magdalenian are the Châtelperronian, about 40,000 years ago; the Aurignacian, 36,500 to 28,000 years ago; the Gravettian, 28,000 to 22,000 years ago; and the Solutrean, 22,000 to 18,000 years ago. (These estimates, and all others in this book, are approximate, and are based on the most current scientific assessments.)

A Mousterian stone axe. Musée Nationale de Préhistoire, Les Eyzies-de-Tayac.
Photograph by Debra G. Aczel.

What, again, was the relationship between these Cro-Magnons and the Neanderthals? There is strong evidence Neanderthals and Cro-Magnons lived in the same territory at the same time—as late as 30,000 years ago. But the Neanderthals became extinct, while the Cro-Magnons flourished and gave rise to modern humans. Some scientists believe that our Cro-Magnon ancestors caused the disappearance of the Neanderthals: either violently, or by fierce competition for resources. As we will see, this replacement of Neanderthals by anatomically modern humans took place in Europe over a relatively short period of time.

The Neanderthals' ancestors are thought to have left Africa hundreds of thousands of years ago and moved into Eurasia. Scientists believe that the Neanderthals and modern humans diverged on two different evolutionary paths long ago.

But then there are the much older fossils of Atapuerca. Finds from a cave here, near the city of Burgos in northern Spain, were first described in a local newspaper in 1863. The complex of caves at Atapuerca has been studied extensively ever since, and hominid fossils that have been dated to 800,000 years ago were found at various excavations late in the twentieth century. This hominid was named *Homo antecessor* (Pioneer Man). Other hominid remains, belonging to the species *Homo heidelbergensis,* were found here, and they exhibit characteristics of both Neanderthal and an earlier hominid, *Homo erectus* (the species to which Peking Man and Java Man belonged). The hominids of Atapuerca were between *Homo erectus* and Neanderthals, and are thought to have been the direct ancestors of the Neanderthals.

Anatomically modern humans likely left Africa around 150,000 to 100,000 years ago. They inhabited Israel (the Qafzeh cave and others) at least 92,000 years ago. They moved into Europe around 40,000 years ago, where for 10,000 years or less they shared the continent with the Neanderthals.

When Dorothy Garrod excavated the cave of Tabun at the bottom of Mount Carmel in 1929, she discovered stone tools that resembled those from the Mousterian industry in Europe, which was associated predominantly with Neanderthals.

Two years after that excavation, Garrod discovered the skeleton of a Neanderthal woman at Tabun. At that time, an associate of Garrod's, the American anthropologist Theodore McCown, who was working at the nearby Skhul cave, found eight burial sites with skeletons. These were of anatomically modern humans. Some researchers have hypothesized that the two caves and their fossils demonstrate that Neanderthals and modern humans interbred in this part of the world. Current DNA analysis of Neanderthal bone marrow may shed light on this hypothesis.

Around the same time Garrod and McCown were working, the French archaeologist René Neuville and his Israeli colleague Moshe Stekelis were excavating at the Qafzeh cave farther east of Mount Carmel, near Nazareth. There they found seven skeletons of anatomically modern humans similar to those from Skhul. Further excavation at Qafzeh revealed additional anatomically modern humans, with distinct chins and modern-shaped skulls. New dating of the Qafzeh cave led to a stunning conclusion: These anatomically modern humans were there some 80,000 to 100,000 years ago. This was announced in 1980 by the Israeli archaeologist Ofer Bar-Yosef, now at Harvard, and his colleague Bernard Vandermeersch. The findings caused great excitement, because they indicated that modern humans occupied the Levant before the Neanderthals. Modern humans, then, did not evolve from Neanderthals.

Later, more accurate dating techniques placed the modern humans of Qafzeh at 92,000 years ago or earlier. The same methods yielded a similar age for the modern humans at Skhul. Anatomically modern humans had occupied the Levant some 30,000 years before the Neanderthals came on the scene. But in Europe, Neanderthals became extinct within several thousand years of the arrival of modern humans.

In 1989, Baruch Arensburg and other researchers published a report in the journal *Nature* describing their discovery of an unusual Neanderthal find from the Kebara cave near Mount Carmel. Among the remains of this individual, which had been buried in the cave, was the fossil of a hyoid bone. This bone, the only one in the body that is not directly attached to the rest of the skeleton, is found above the larynx, or voice box, and in humans it plays an important part in speech. On the basis of this finding, Arensburg and his colleagues argued that the Neanderthals may have possessed language.

Ian Tattersall of the American Museum of Natural History in New York is skeptical about claims that the Neanderthals possessed symbolic language. He maintains that while they may have had a system of vocal communication, the fact that the Neanderthals left us no cave art, or art in any form—as did the Cro-Magnons with their magnificent cave paintings—proves that Neanderthals had not developed symbolic thinking. Tattersall sees the ability to think symbolically as the pinnacle of development of the human mind. It is this "Great Leap Forward," as others have called it, that includes the important development of language.

Leaving aside the probability that the Neanderthals may have been able to produce some grunts and express some syllables, Tattersall and his supporters think that a full language was beyond Neanderthal ability, even though the average Neanderthal brain size was greater than ours. These scientists believe that something other than size was the determining factor, and that this element—unique to *Homo sapiens*—produced language and symbols. The Neanderthals, according to these experts, had not made the leap to symbolic thinking. And perhaps therein can be found the reason for Cro-Magnon's survival and Neanderthal's ultimate demise.

Other scientists disagree. Donald Johanson says this about Neanderthal speech: "Whether or not Neandertals possessed language, finally, remains a matter of debate. Although they may not have been fully equipped for modern language, it seems unlikely to me that Neandertals sat silently around the campfire in their caves."

In the February 23, 2006, issue of *Nature*, an article by Paul Mellars of the department of archaeology at Cambridge shed light on the mystery of the disappearance of the Neanderthals. Mellars's article, "A New Radiocarbon Revolution and the Dispersal of Modern Humans in Eurasia," made use of a new calibration system. What the adjustment did was shift the dates back by several thousand years.

One result of this correction was that the scientifically estimated dates for the arrival of anatomically modern humans in Europe and their dispersal over the continent became sharper. According to the new analysis, anatomically modern humans swept through Europe much faster than had previously been estimated: They virtually took over the continent, moving at a rate of a third of a mile per year, and reached their farthest places of habitation, on the Iberian Peninsula, within 5,000 years of their arrival on the outskirts of Europe. According to this analysis, it took modern humans from about 46,000 until about 41,000 years ago to take over Europe. This rate of movement is similar to the rate of expansion of agricultural communities, which began in the Near East about 11,000 years ago. The study has confirmed earlier results from Ofer Bar-Yosef about the speed with which anatomically modern humans occupied Europe.

While new dates for Neanderthals should also be obtained by recalibration, it is already evident that the anatomically modern humans replaced the European Neanderthals much faster than had been thought. In France, for example, Mellars's analysis implies that Cro-Magnons inhabited the same regions as Neanderthals for not longer than a thousand or two thousand years. The Neanderthals' last stand seems to have been near Gibraltar, on the southern tip of the Iberian Peninsula, where Mousterian tools—believed to have been made by Neanderthals—have been found among relics dated to about 30,000 years ago. This was the latest (still unconfirmed) Neanderthal date, and it is currently not believed that Neanderthals survived beyond 30,000 years before our time. Overall, the estimated dates indicate that Neanderthals became extinct within some 10,000 years of the completed takeover of the continent by anatomically modern humans. Other analyses narrow this range to as low as 6,000 years of joint occupation of the continent.

Why did the Neanderthals disappear, while anatomically modern humans took over? This is the Neanderthal enigma. One possibility is abrupt changes in the environment. The world was seeing sharp temperature changes as the glacial era gave way to warming spells. The Neanderthals may not have adapted well to the new, warmer environment. They also had competitors for resources. And they may have been killed off. Ofer Bar-Yosef believes that the anatomically modern humans' routes of migration into western Europe split the Neanderthal habitat, thus splintering Neanderthal communities into groups of fewer than four hundred individuals each. Such small communities do not possess enough genetic diversity for long-term survival.

In the early part of the twentieth century, some scientists believed that Neanderthals were our direct ancestors. In 1908, Neanderthal fossils were discovered at the cave of La Chapelle-aux-Saints in southwestern France. The individual whose bones these were was given the nickname "Old Man," and the remains were examined by French archaeologists, who proposed that the Neanderthals were our ancestors.

The Chapelle-aux-Saints fossils were discovered by three French clerics, and on the advice of the Abbé Breuil were sent for analysis to the French paleontologist Marcellin Boule at the Museum of Natural History in Paris. Boule classified the fossils as *Homo neanderthalensis,* stressing that this was not a modern human. He compared the skull of the Neanderthal with the skull of a chimpanzee and that of a modern human in making this determination.

Boule conducted an exhaustive analysis of the Chapelle-aux-Saints Neanderthal. He made a plaster cast of the inside of the skull and, on the basis of the relative sizes of different parts of a modern brain, deduced that the Neanderthal had "rudimentary intellectual faculties." He concluded that even though the brain itself was relatively large, the intellectual capacity of the Neanderthal was low because of the distribution of the mass of brain tissue, which was different from that of modern humans (because the Neanderthal skull was lower and longer than ours). He also noted differences in presumed posture and in the size of the legs, and concluded that Neanderthals were shorter than modern humans.

The Neanderthal mystery is multifarious. Who were these people? *Were* they people? Could they speak? What did they use their large brains for? How did they interact with the anatomically modern humans they encountered (if indeed they met)? And why did they disappear so quickly—faster than many other animal species that have gone extinct—as modern humans swept through Europe?

In 2006, the Max Planck Institute for Evolutionary Anthropology in Leipzig, Germany, together with the American company 454 Life Sciences, based in Branford, Connecticut, announced a plan to reconstruct the Neanderthal genome. Work on the genome is proceeding. The Swedish scientist Svante Pääbo, working at the Max Planck Institute, has successfully drilled holes in fossil Neanderthal bones and extracted the remains of bone marrow. These samples are subjected to careful procedures that can retrieve very small pieces of the DNA code, which are then painstakingly reconstructed by the 454 Life Sciences staff to yield genetic information. Once this genome is complete, it could be compared with that of modern humans and, scientists hope, could provide us with more information about the possible genetic relationships between the two species.

Pääbo and his colleagues first extracted Neanderthal mitochondrial DNA as early as 1997, and in 2000, Igor Ovchinnikov and colleagues succeeded in extracting DNA from the remains of a Neanderthal baby found in the Mezmaiskaya cave in the Caucasus. In the years since, this science has developed and we are approaching the point of uncovering the Neanderthal genome.

One human gene believed to play a role in the development of language is labeled FOXP2. Geneticists working on the reconstruction of the Neanderthal genome hope to find the equivalent of human FOXP2 in their samples of Neanderthal genetic material and compare it with ours. Such an analysis should help scientists assess the chances that the Neanderthals possessed a language.

Many scientists believe that it would not be unusual for several hominid species to have coexisted; the oddity is rather that we are now the only human species alive today. Coexistence of related species seems to be the rule in nature, not the exception. And if research findings about a fossil discovery made in Java in 1996 are correct, then not two, but *three* different human species have inhabited our planet *at the same time* as late as 30,000 years ago. These were *Homo sapiens*, *Homo neanderthalensis*, and *Homo erectus*—the remnants of a hardy species of human ancestors that had widely inhabited our planet for well over a million years.

Chapter 5

JAVA MAN

The third hominid to be discovered by science was *Homo erectus*, the first specimen of which was unearthed in 1891 by Eugène Dubois, who spent years digging for fossils of human ancestors on the islands of present-day Indonesia.

Dubois was born on January 28, 1858, almost a year before Darwin published *On the Origin of Species*. Dubois grew up in Eijsden, in southern Holland, near the Belgian border. His family was Catholic, religious, and conservative; one of his sisters became a nun. Eugène's father was a pharmacist and at times served as mayor of the town. The family was wealthy, and one of the better-educated in the region. As a boy, Dubois was restless and hardworking. Tall, fair-haired, and handsome, he became an excellent swimmer, often disappearing from his family's large house to swim in the nearby river. An exceptionally bright student, he showed the most promise in his study of natural science.

In childhood, Dubois heard about the Neanderthal discoveries in Gibraltar and Germany, and was enthralled. He read whatever he could about evolution, and studied about it in school. When he was ten years old, a German scientist named Karl Vogt came to a nearby town to lecture on evolution—the new theory of the day—causing excitement and sparking controversy.

Eugène had to content himself with reading newspaper reports about Dr. Vogt's lecture, which deeply shook his audience. He claimed that man was a cousin of the apes, and that the creatures on our planet were related to one another through common descent. By contrast, his listeners believed that the world was static and rejected the idea of change. But Eugène Dubois was so taken with the theory of evolution that he kept searching for more information about it. He understood the relationship between fossils and evolution, and knew that Neanderthal was an extinct form of man, yet not far from modern humans. Therefore, Neanderthal's importance was limited—he was too close to us. What Dubois wanted to do from a young age was find the actual link, the missing link between apes and man. This would be a hominid that was, in some sense that could be discerned from fossilized remains, halfway between modern humans (along with Neanderthals) and apes. Dubois's mission in life was to show that Darwin was right, by providing evidence of that link—proving that man descended from an apelike ancestor.

With this aim, Dubois registered to study medicine at the University of Amsterdam in 1877. He became a student of the botanist and evolutionist Hugo de Vries, and after completing his studies in 1881, was immediately offered two competing positions at the university. He took the first offer: an assistantship in anatomy with Professor Max Fürbringer.

Dubois began his work in earnest. He was an able researcher, though not a very popular lecturer. His gifts lay in his ability to make discoveries in anatomy, not in educating young doctors. In 1884 he qualified as a physician, and soon started a comprehensive study of the structure of the larynx. Dubois was able to establish the evolutionary path from gills to larynx as a speech organ. His goal was to solve one of the greatest mysteries in evolution: the development of speech, and the discovery about the anatomy and evolution of the larynx was an important step in this direction.

But when Dubois showed the draft of his research paper to his professor, Fürbringer told him that he had derived the idea by listening to him, and that therefore the concept must be acknowledged as partly due to him. Dubois was flabbergasted—he was angry and stunned by this unreasonable demand. Though he eventually did as his professor had asked, the residual resentment would push him away from academia.

In 1886, Max Lohest, a Belgian geologist, presented his analysis of a new discovery: a third Neanderthal, found at Betche-aux-Roches in the region of Spy, Belgium. The finding invigorated Dubois's quest for proof of human evolution. While the Neanderthals were not old enough to be the missing link between humans and apes, somewhere in the world must be buried the remains of that link.

Finding the fossilized bones of hominids is an exceptionally daunting task. It is like the proverbial search for a needle in a haystack. A hundred fifty years after the discovery in the Neander Valley, we still have precious few fossils, despite the fact that so many people, using ever-advancing technology, have been spending decades in search of them. It is with this perspective that we must judge the audacity of a young Dutch doctor who resolved—at a time when very few hominid fossils were known—that he would find the missing link.

In keeping with his training as a scientist, Dubois went about preparing his search in a logical, systematic way. Apes live in the tropics, Dubois knew, so the tropics had to be where our common ancestor with the apes lived. But where? In Dubois's day it wasn't known that the chimpanzee was our closest living relative. Some evolutionists believed that gibbons were, and Dubois favored this theory. Since gibbons live in Asia, he concluded that Asia was the right place to look for the missing link. The continent is huge, and even its tropical region—where apes live—is quite large.

Dubois had heard that fossils of great apes had been discovered in India and Java. Since India was British and Java was a colony of his own country, clearly it would be easier for him to go to Java or some other island in the Dutch East Indies. He found additional guidance in *The Geographical Distribution of Animals*, by Alfred Russel Wallace, Darwin's fellow naturalist, which appeared in 1876. Wallace defined an invisible line passing between the islands of Bali and Lombok in the Indian Ocean. While Bali and Lombok lie very close to each other, and are both among the Lesser Sunda Islands of the seemingly uniform Malay Archipelago, Wallace had concluded that they belonged to two different geographical regions: Asia and Australia. He determined that islands to the west of his line constituted part of Asia, while the islands east of the line were part of Australia. And indeed, marsupials—typically Australian animals—were found to live only east of the Wallace Line, while placental mammals lived to the west of it. Dubois understood that in order to find the missing link he must look only where great-ape ancestors must have lived, that is, on the Asian side of the Wallace Line. This left him Bali, Java, Sumatra, and several smaller islands of the Dutch East Indies.

Now Dubois faced the hardest task: telling his wife, his parents and in-laws, and his employer and colleagues at the university about his decision. To his surprise, his wife, Anna, supported him in this decision. She would have to give up Amsterdam, and all its comforts, as well as friends, family, social life, and culture. Both their parents were vehemently opposed to the strange decision and tried to make Dubois change his mind. At the university, where he had some close friends and colleagues who cared much about him and recognized his excellent prospects for advancement to professor, there was a strong effort to make him stay. But he was dedicated to the idea of finding the missing link in the Dutch East Indies.

Dubois tried to convince the Dutch government that financing a scientific expedition to the East Indies to search for proof of evolution was a good idea. But his pleas fell on deaf ears, partly because the conservative and religious authorities were not interested in the missing link or in evolution. So Dubois decided to use his training as a doctor to get him to the island colony. The Dutch military had a strong presence there, and armies always need doctors. Dubois signed up for an eight-year stint with the army, assigned to the East Indies. In October 1887, the Dubois family—Eugène, Anna, and their daughter, Eugénie—left Amsterdam aboard SS *Prinses Amalia*, headed for the East.

Surviving the hardships of the tropics in Sumatra, the family lived on an army base for several months. Another child was born while Eugène worked at the military hospital, treating everything from gunshot wounds to malaria and typhoid. On his days off, he ranged the countryside looking for caves.

After about a year of searching mountain caves, Dubois found some fossils. These were not hominid remains, but bones of extinct animals that had lived at about the time he thought the hominids should have been around. These animals included cave bears, rhinoceros, and elephants. Buoyed by these discoveries, he wrote to the Dutch governor of the colony, emphasizing the importance of looking for fossils; he pointed out that if the Dutch wouldn't search for them, the British would. And once they found them, the British would take away scientific glory that was rightfully Dutch. The governor responded positively and even gave Dubois a number of workers to help him in his search. These were peasants who owed taxes and would be forced to fulfill their obligations by several weeks of labor.

Dubois was able to parlay his early success into a wider-ranging operation, which took him to caves in a broad swath of the mountains of western Sumatra. His retrieval of animal fossils and his subsequent reports on his finds, which he sent to the governor and to the authorities in Holland, won him release from hospital duties and an increase in the number of forced laborers assigned to his project.

With as many as fifty laborers and convicts, Dubois continued his search with renewed vigor. He would leave his wife and children in town for long periods to explore caves, his laborers excavating these caves in search of fossils. After a peak in the number of animal fossils found, in 1888, the next year brought him few new finds and a growing frustration with the elusive missing link. In addition, he succumbed to malaria, and the bouts of fever and chills left him weak and demoralized.

After his recovery, Dubois started rethinking his strategy. The Sumatran caves had yielded some success, but by now he had been excavating them for more than two years. It was time to start looking elsewhere. The previous year, he had been shown a skull that had been found in 1888 at Wadjak in neighboring Java. The skull belonged to a human, not an early hominid, although it had some unusual features. Dubois decided that Java might yield earlier skulls—perhaps even the missing link.

In 1890, Dubois again uprooted his family and moved to Java. But Java had a different geography from Sumatra's. There were only flat plains through which rivers flowed, and high, steep mountains—caves would not have been the place to look, since early humans likely did not inhabit the steep mountains; riverbeds offered a more promising terrain. Those years of hunting for fossils had taught Dubois where to look, and he chose a location along the bank of the Solo River near the village of Trinil. Here he embarked on an intensive excavation operation. He now had two corporals from the Dutch engineering corps supervising fifty laborers digging along the riverbank.

In 1891, the excavators began to recover fossils. The first were the remains of extinct animals similar to those Dubois had found in Sumatra. But as the work continued, the laborers brought him teeth that looked human. This was an exciting find, and the digging intensified in pace and extent. Then, in October 1891, Dubois's workers recovered a flat skullcap, one that was lower than those of the Neanderthals found in Europe. Dubois knew at once that this was what he had been looking for—a link between humans and apes.

These fossils, which became known as Java Man, are now identified as belonging to the species *Homo erectus*, an ancestor of modern humans. Java Man lived about 700,000 years ago, making him the oldest of the hominid finds recovered by 1891. Java Man was an exceptional fossil because he was more primitive in appearance than the Neanderthals, and yet more advanced—in cranial capacity and other features—than the apes. The skull was more humanlike than those of apes, but the teeth were apelike.

Dubois returned to Holland and spent the rest of his life fighting for recognition for his fossil, which he named *Pithecanthropus* (Ape Man). He believed that his *Pithecanthropus* was the missing link—it was neither human nor ape, but rather something in between. Yet the scientific community in Europe failed to award him recognition for his achievement, and he became increasingly frustrated. He hid his find under the floorboards of his house in Holland and refused to show it to anyone.

Years later, when people came to see him, hoping to inspect the fossils, they would be told that Dubois declined to see anyone, and that he had his fossils hidden away. Pat Shipman reports in her fascinating biography of Dubois that people would often be told: "Do not be angered if he will not see you. He is not in the habit of receiving visitors these days, especially those who want to see his fossils." Only years later did Dubois allow the fossils to be moved to a museum.

When Peking Man was discovered years later, Dubois fought hard for a distinction between the two hominids. But science eventually determined them both to be members of the species *Homo erectus*, Upright-Walking Man. Dubois's claim to have found a species between humans and apes was confirmed by the discovery of Peking Man.

Dubois's work in Java was continued by the German paleontologist G. H. R. von Koenigswald. He visited Dubois and tried to see his fossils, but Dubois would not show even him the skull of Java Man. Koenigswald, however, still believed that Java Man was the "most famous, most discussed, most maligned fossil" ever found. After the fossils were locked away in the safe of a museum in Leyden, Koenigswald was able to see them. He then went to Java to search for further Java Man remains to prove Dubois right.

Koenigswald made discoveries of fossil hominids in Java, but Dubois would not recognize these as belonging to the genus *Pithecanthropus*. Years later, Koenigswald would be visited by Pierre Teilhard de Chardin, who would examine his finds with much interest and with him look for new ones.

Chapter 6

TEILHARD

The philosopher, theologian, geologist, paleontologist, and mystic Pierre Teilhard de Chardin was born on May 1, 1881, in his family's eighteenth-century château, named Sarcenat, some two miles from the village of Orcines and four miles east of the city of Clermont-Ferrand, in the Auvergne region of central France. He was fourth of the eleven children of a wealthy family; his father's ancestors had lived for centuries in châteaux in the hills of the Auvergne. Sarcenat, the family's principal residence, is situated on high ground, and an elegant avenue of chestnuts and elms leads to its gate. The stately mansion, with its tall windows and gray shutters, flanked by white turrets, commands majestic views of the conical extinct volcanoes of Puy-de-Dôme in the distance, the rolling hills nearby, and the sprawling suburbs of Clermont-Ferrand to the west.

This region of France has been called a "geologist's paradise" because of its varied terrain, from flat plateaus to towering volcanoes to steep ravines. The members of the family enjoyed their country living, and Pierre's father, Emmanuel, was an avid collector who was interested in geology and in natural science in general—he collected stones, insects, plants, and anything else he could find in the countryside surrounding his estate. He inspired his son's growing curiosity about the natural world.

Emmanuel was a descendant of Pierre Teilhard, who in 1325 was given noble rank by Charles IV for his services as notary to the French crown. Two hundred years later, in 1538, his descendant Astorg Teillard was also ennobled. Another title of nobility was awarded by Louis XVIII in 1816 to a more recent ancestor, also named Pierre Teilhard. In 1841, this Pierre's grandson, Pierre-Cirice Teilhard, married a noble-woman, Marguerite-Victoire Barron de Chardin, and the two names, Teilhard and de Chardin, were united. The son of Pierre-Cirice and Marguerite Teilhard de Chardin, born in 1844, was Emmanuel, our Pierre Teilhard de Chardin's father.

Pierre's mother, Berthe-Adèle de Dompierre d'Hornoy, was born in Picardy, in northern France, and was a great-grandniece of Voltaire. She was a deeply religious woman, who woke up before dawn every morning and walked several kilometers to attend Mass. She encour-aged her son to pursue a spiritual life, and under her tutelage the boy followed religious training. In 1893, at the age of eleven, he enrolled at the Jesuit school of Notre Dame de Mongré in the town of Ville-franche-sur-Saône. He studied there for five years and was consistently at the top of his class.

The large Teilhard de Chardin family was close to a related family named Teillard-Chambon. The Teillard-Chambons owned a mansion in Clermont-Ferrand, and the two families met regularly. Through these family gatherings, Pierre developed a friendship with one of his cousins, Marguerite Teillard-Chambon, who would remain his confi-dante throughout his life. She later became a writer and chose the pen name Claude Aragonnès, under which she published collections of Teilhard's letters.

Even as a child, Pierre Teilhard de Chardin had an amazing aptitude for science, but he was equally devout in his religious practice. He pos-sessed a unique mixture of qualities and interests that made him suited for both science and faith, the one area encouraged by his father, the other by his mother. In each of the five years he spent at the Jesuit school, he earned distinction for religious devotion, and at the same time won great praise for his achievements in science. It was at the school that he became an ascetic who voluntarily rose at dawn every day and went to sit in the chapel, often in freezing temperatures, to read religious writings before the rest of the students awoke. He would follow a similar habit throughout his life, wherever he might be: in an Asian desert, in a prehistoric cave, or aboard a ship in rough seas.

Teilhard found the life of a Jesuit attractive and fulfilling. On June 4, 1897, just before graduation, he wrote a letter to his parents telling them that he wanted to become a Jesuit priest. After graduation in August, he spent more than a year at home in Sarcenat, at the request of his parents, who thought he was physically weak and needed strengthening. On March 20, 1899, he moved to Aix-en-Provence, where the Jesuit novitiate for the province of Lyon was located in a somber gray building attached to a church on the quiet Rue La Cépède, to begin his training as a Jesuit. Teilhard was described by his fellow novices as modest, shy, loyal, always ready to help, and anxious not to be perceived as different from others. He wanted to fit in. He was tall and a good walker, and had a ready smile. The strenuous program of study, prayer, and meditation left no time for exploring the geology of Provence, something he would have loved to do.

In October 1900, Teilhard moved from the Jesuit novitiate in Aix to the juniorate in Laval, in northwestern France. He lived at the Maison Saint-Michel, a large building complex owned by the Jesuits, and there, on March 25, 1901, he took his first vows in the Society of Jesus. That day, he wrote to his parents: "At last I am a Jesuit: I don't have the time to write you at length today, but I want to tell you about my happiness at being finally entirely at the Sacred Heart near the Saintly Virgin. If only you knew the joy I feel now that I have at last given myself completely and forever to the Society, above all at a time when it is being persecuted."

Indeed, the Jesuits were being persecuted in France, and religious communities were being secularized by governmental decree. In 1901, the country, still reeling from the Dreyfus Affair, which rocked the nation and brought out the specter of anti-Semitism, focused its attention on the separation of church and state. French lawmakers considered passing laws designed to remove clerics from state positions in education. Premier René Waldeck-Rousseau and his successor, Émile Combes, then pushed through and implemented these laws. The legislation hurt the Jesuits, since they had been the order most involved in instruction at religious schools in France, and it prompted a mass exodus of Jesuits from the country.

In fall 1902, Teilhard, like many fellow Jesuits, hastily left France for Jersey, a British crown dependency and one of the Channel Islands, to continue his training. The Jesuits had an infrastructure there, consisting of buildings and personnel, so that young Jesuits could continue their education. Teilhard spent three years on Jersey, studying Greek and Latin classics and scholastic philosophy.

In 1906, the Jesuits sent Teilhard to teach at their school in Cairo. On his arrival in Egypt, he was fascinated to see minarets all around him, hear the muezzins' calls to prayer, and beyond, view the imposing pyramids at Giza. On weekends he ventured into the surrounding desert in search of stones, shells, and fossils. He made friends at the Egyptian Museum, and these new connections allowed him to devote time to poring over interesting fossils that had been recovered from Egypt's Western Desert.

He observed tortoises along the Nile, deadly vipers (one of which he caught and preserved in alcohol), lizards, rats, and insects. At the school he taught physics, which had attracted him from an early age and which he had studied while pursuing his religious training. Teilhard felt the Orient beckon him, and later in life he would recall fondly his first experience in exotic Egypt. This assignment, along with stories he had heard from his siblings who had traveled widely, made him seek travel opportunities throughout his life.

Three years after his arrival in Cairo, Teilhard was ordered to return to Europe, this time to Hastings, East Sussex, on the southeast coast of England, where Jesuit exiles from France had established themselves. Teilhard studied theology, and also spent days searching for natural relics he could add to his collection; but now he extended his interest to living things, the flora and fauna of the English countryside.

At about the same time, Teilhard was reading a book on evolution by the influential French philosopher Henri Bergson, *Creative Evolution*—a work the Vatican would place on its Index of Forbidden Books. The ideas in the book spurred Teilhard to learn more about the theory of evolution, for he found a scientific justification for the unity he felt he shared as a human being with the entire world of living creatures. It was at Hastings that Teilhard decided to devote his life to the pursuit of two distinct yet parallel tracks: science and religion. He began to examine the process he believed was evident everywhere in nature—the progression of matter into living organisms and the evolution of these organisms to ever more complex life forms.

Teilhard had already moved in the direction of consolidating elements of science with those of religion when he was a boy. Walking in the hills of his native Auvergne, he collected old discarded items made of iron and saw in them a symbol of strength. When he later realized that iron rusts, he concluded that everything in nature—even solid, inanimate objects—undergoes change. He started to develop a concept he called "the All" (*le Tout*)—the totality of the entire universe—and a passion for the changeability, or constant evolution, he saw in this All. In Egypt, collecting fossils in the desert, he went further in this direction, believing that an evolutionary force brought constant change to living things. And in Hastings, as he read Bergson's book, the ideas of evolution became so powerful that they convinced him that everything in the universe, inanimate objects and living systems alike, was in constant flux, ever evolving as decreed by God. The goal was a point where everything would converge to form the body of Christ. This was Teilhard's Omega Point.

Teilhard's embrace of science and religion was deep and complete. The roles played by these two seemingly contradictory elements were symbolized by the two images that he displayed next to his bed throughout his adult life: one of Christ, and the other of Galileo. Where others may have seen a conflict, for his entire long career, Teilhard never had the slightest difficulty reconciling science with faith. Since he grew up with the twin routines of exploring the natural world around him and worshipping God, science and religion were a unified entity in his mind. To him, many stories in the Bible were allegories, and the true history of the planet was written in rocks and minerals and fossils.

The chalk cliffs of Hastings were a popular location for amateur paleontologists, who liked to explore this terrain suited for the preservation of ancient bones. Because he spent so much of his free time digging around the grounds of the seminary, Teilhard attracted the attention of local natural scientists, including the director of the Hastings Museum (which now houses both paleontological displays and an art gallery). Another amateur paleontologist whom Teilhard befriended was Charles Dawson, known as "The Wizard of Sussex," because of his many fossil discoveries. Dawson later claimed to have found Piltdown Man, proven in the 1950s to have been a hoax.

Pierre Teilhard de Chardin was ordained a priest on August 24, 1911. In July of the following year he returned to France, taking up residence within the returning Jesuit community around the Rue de Rennes in Paris, in the heart of the Left Bank. Like many other Jesuits, he pursued a rigorous intellectual education, specializing in science and philosophy. He signed up to study geology with Georges Boussac at the Institut Catholique, and paleontology with the famous scientist Marcellin Boule at the Paris Museum of Natural History.

Teilhard first met Boule in his laboratory. He was somewhat apprehensive about approaching him, since he had heard so much about the paleontologist's gruffness and apparent unfriendliness. He also knew that he was a sworn agnostic, and this worried Teilhard even more. But Boule was charmed by him. They learned that they came from the same part of France, and this helped cement a lasting friendship between them. Boule agreed to take Teilhard as a student, and assigned him the task of classifying a large collection of fossils that had recently been discovered near Toulouse. The anti-religious professor and the devout Jesuit student respected each other's talents and wisdom.

Yet Teilhard's work at the museum was to be interrupted. In 1914, while on a mountain-climbing trip, he heard the news that France had entered the war. He felt an urge to defend his country, but no order calling him to arms arrived. By the end of August, the Germans had broken through the defenses of Belgium, reached France, and were within thirty miles of Paris. The Jesuits, by then reestablished in France, decided to send Teilhard to England to finish his religious training. Soon after he went to England, his call-up papers arrived, and he returned to France and the Auvergne. From there, he went to join his army unit.

Teilhard wanted to see action, but as a priest he was prohibited from taking part in fighting. He was thus assigned to be a medic/stretcher-bearer, attached to the Eighth Regiment of North African Zouaves. His unit was sent first to the Marne, where the French had been making progress repelling German attacks, and then to Ypres, in Belgium.

It was in Belgium that Teilhard experienced the true horror of World War I. When they arrived at Ypres, the troops found a town that had just been burned down. Hundreds of soldiers lay on the ground, dead or dying. And after the Germans were through with their conventional-weapons strike, they attacked their enemy with poison gas.

Teilhard survived these hellish onslaughts, saving many of the wounded. But his soul was scarred for life. There, and in Artois in France, where they were subsequently moved, the Zouaves lost half their men. At his own request, Teilhard was moved from his post in the back tending the wounded to a forward position in the trenches. The divisional order mentioned him: "He displayed the greatest self-abnegation and an absolute contempt for danger." In 1916, the unit was moved back to Belgium. Teilhard went through a difficult period in the face of the unbearable horrors of war. In addition to his medical duties, he ministered spiritually to the troops in the battlefield. On June 20, 1917, he was awarded a military medal for his heroism, which included risking his life by approaching within a dozen yards of the German lines to rescue the wounded, and going alone into trenches "under violent bombardment to recover a wounded soldier," as his citation read.

While on the front, and in an effort to reconcile the atrocities he observed with his belief in a loving God, Teilhard wrote an essay that attempted to put these horrific experiences in perspective. He titled it "La Vie Cosmique" ("The Cosmic Life"), and it contained the elements of his developing philosophy and his mysticism. "What follows springs from an exuberance of life and a yearning to live," it began.

It is written to express an impassioned vision of the earth, and in an attempt to find a solution for the doubts that beset my action—because I love the universe, its energies, its secrets, and its hopes, and because at the same time I am dedicated to God, the Origin, the only Issue and the only Term. . . . All around us, in whatever direction we look, there are both links and currents. Countless forces of determination hold us in their grip, a vast heritage from the past weighs down upon our present, the thousand and one affinities by which we are influenced pull us away from ourselves and drag us towards an end of which we have no knowledge.

The essay was rich in images, including the multitudes of the universe: grains of sand, and stars in the heavens; God permeating the ether; the good earth, and people's communion with it; pain and pleasure and hell; the struggle with the angel; and finally, evolution.

Teilhard sent this article to the Jesuit periodical *Etudes* in Paris. But when its editors read it, they were disturbed by the bizarre images in Teilhard's text, which they interpreted as pantheistic and out of line with Catholic thought. Jesuit leaders began to suspect that Teilhard was not following the life trajectory expected of a Jesuit. His writings evinced a mind that was both unconstrained by conventional religious thought and deeply affected by the severe pain brought on by the experiences of war, which manifested themselves in the disturbing images in his text. The charge of pantheism would plague Teilhard throughout his life. But perhaps the Jesuits were equally worried about the appearance in the essay of the concept of evolution, which he couched within a Christian framework, writing: "The life of Christ mingles with the life-blood of evolution."

After lengthy consideration, the editors rejected Teilhard's essay. One of them, Father Léonce de Grandmaison, explained to him: "Your thesis is *exciting* [he used the English word, in the midst of French] and interesting to a high degree. . . . It's a rich canvas, full of beautiful images. But it is not at all suited for our peaceful readers." Teilhard was disappointed, and even wrote his cousin Marguerite Teillard-Chambon that if this was any indication, his ideas would never see the light of day. He was eager to publish, and without the possibility of publication, he lamented, his ideas could be disseminated only "by conversation, or as manuscripts passed under a coat."

This was a prescient assessment. As his church forbade the publication of his ideas over the coming decades, Teilhard allowed their dissemination as typed pamphlets passed around among his many friends—in some cases, indeed under a coat, as he predicted. Teilhard understood from the start that, since he was a Jesuit, his writings would have to pass through levels of ecclesiastical censorship: those of his own order in Rome, and those of the inquisitors of the Holy Office of the Vatican, delegated to protect faith and morals, who could threaten to place a work on the Index of Forbidden Books in order to suppress the dissemination of subversive ideas. But Teilhard was reconciled to this scrutiny and accepted it as the right of the Church.

Teilhard admitted soon after writing "La Vie Cosmique" that he had "a naturally pantheistic soul." In his own mind he reconciled this idea of a multitude and a "one" with a deep and sincere Christian belief. Teilhard's pantheistic feelings made his provincial in Lyon, Father Claude Chanteur, nervous about allowing him to take his solemn vows in the Society of Jesus. But another Jesuit, Father Vulliez-Serment, convinced the provincial that Teilhard was a solid Jesuit, and on May 26, 1918, at Sainte-Foy-lès-Lyon, while on leave from the front, Pierre Teilhard de Chardin took these vows, which included chastity, obedience, and poverty. He was now fully committed to the Society of Jesus. And yet, even at that moment of total submission to the Church and its rules, he continued to interpret his role as a priest in his own way. On July 8, while stationed with his unit in the forests along the Aisne River near Compiègne, northeast of Paris, he wrote: "This is why I've dressed my vows, my priesthood (and therein my strength and my joy), in a spirit of acceptance and deification of the Powers of the Earth."

Having returned to Paris after the War, Teilhard plunged into his studies and finished his thesis at the Museum of Natural History. He took courses at the Sorbonne in botany, zoology, geology, and paleontology, and continued to write essays on how matter in the universe was transformed into living organisms, pursuing this track to evolution. In his own mind, he reconciled evolutionary ideas with his religious faith by positing the concept of a dynamic Earth, propelled by God toward Him and the ideal of divine perfection.

Teilhard was a gifted paleontologist, professionally groomed by Boule for a career in the field. He was learning much about science and about prehistoric fossils: how to find them, how to clean and prepare them for analysis, and—most important—how to decipher the stories these ancient relics had to tell. And this scientific work moved him ever forward in the direction of evolution. Studying bones from various geological periods intensified his notion that our universe moves forward to more advanced, more complex states. With his equal enthusiasm for physics and astronomy, he understood evolution in the same way he knew our planet to be in motion, as Galileo had argued, using the ideas of Copernicus. Teilhard believed that Earth was also a progressing biological system, eventually reaching cognition in humans, and continuing toward hyperconsciousness: the consciousness of the entire planet.

Botany taught Teilhard about the natural forces that promote vegetable growth, and about the connections among various forms of life made through molecular processes. Zoology showed him that there was an immense variety of living organisms, dating back many millions of years, and progressing toward complex systems such as those in the human body. One of the living organisms whose remnants he studied held a special place for him. It was a 50 million-year-old palm-sized primitive primate with large eyes and long legs, whose fossilized remains had been found a few years earlier in Cernay, in the fertile Rhine Valley in eastern France, and brought to the Museum of Natural History as part of the collection of Victor Lemoine, a noted flower-breeder.

As TEILHARD MATURED, he remained deeply religious. He performed his meditations and prayer with great devotion, and continued to believe in God and maintain his loyalty to his order. Yet he believed that the Catholic Church and the Jesuits needed a better understanding of what they were saying about God. One point of contention was the story of the spontaneous appearance of Adam and Eve in the Garden of Eden, which Teilhard refused to accept literally. He knew that the advent of humans was through the gradual process of evolution. But this contradiction never presented him any problems of belief. Much of scripture should be taken allegorically rather than literally, he knew, and he saw no conflict between embracing evolution and at the same time practicing his religious belief as a devout priest. In the same way that physical laws explained the physical universe, evolution was the explanation of the arrival of human beings.

In 1920, when Teilhard was thirty-nine, he was appointed chair of geology at the Institut Catholique. He lectured there with a strong sense of mission. The following year, on May 21, 1921, at the request of members of his wartime regiment, he was awarded the rank of *chevalier*, or knight, in the French Legion of Honor. His citation read: "An outstanding stretcher-bearer, who during four years of active service was in every battle and engagement the regiment took part in, applying to remain in the ranks in order that he might be with the men whose dangers and hardships he constantly shared."

In December of the same year, Teilhard's thesis was published in the journal *Annales Scientifiques*, and in the following March he was awarded a doctorate in science from the Sorbonne; this gave his appointment at the Institut Catholique even greater authority. His thesis, an analysis of fossilized mammal bones from the Lower Eocene period (about 50 million years ago) found in France, compared cranial capacities of various extinct animals and argued that brain size increased over time. Teilhard defended his thesis before a panel of some of the greatest French paleontologists and an overflowing audience.

Since his thesis had already been published—as good dissertations in France often are—and had been well received by the academic community, whose members commented on it and nominated it for an award from the Geological Society of France, its defense was much easier than the defense of theses by candidates whose work had yet to be recognized. Still, because he was a visibly religious man, some of the scientists present at his defense wondered whether a man of faith could be open-minded and objective about matters of science, as a good researcher should be. Their questions, therefore, were at times pointed. Teilhard answered them all so well that after a half-hour of interrogation the head of the academic jury commended him for his "clarity of spirit and professional gifts." His degree was awarded "with highest honors." The forty-year-old Jesuit was becoming a major French intellectual and scholar.

With his new fame came invitations to lecture. One request was from a theological institute in Enghien, Belgium, where Teilhard traveled in the spring of 1922. He delivered a lecture in which he presented his view of the evolution of mammals, based in part on his analysis of French paleontological finds discussed in his thesis. He also expounded his position on original sin, expressing his view that the biblical Adam and Eve could not have existed as described in Genesis, since one couple could not have led to the entire human race. Similarly, he expressed skepticism about the idea of the terrestrial paradise. These notions simply did not stand up under the scrutiny of science. Teilhard suggested, instead, that the Fall of Adam and Eve as described in Genesis must be considered a shorthand description of all the human failings, infidelity, and cruelty we see in this world.

Teilhard left Belgium contented, but some weeks later sensed that his ideas about original sin were not fully developed. He expanded them, and wrote a working paper titled "A Note on Certain Historical Representations of Original Sin," in which he built a theological framework for discarding the literal idea of original sin. Quoting Saint Paul in his Epistle to the Romans 8:22, in which he described a world that "groaneth and travaileth in pain together until now," Teilhard suggested that Paul was implying that human sinfulness, pain, and death were the necessary shadow of the redemption.

But Teilhard wasn't happy with this approach, either; the cataclysmic nature of the Fall, he thought, could turn people away from religion. He continued to develop his paper aimed at rejecting the literal interpretation of the Fall of Adam and Eve. He kept this paper in an unlocked drawer in his office at the Institut Catholique, not realizing the grave error he was committing by writing these things down and by leaving the paper in the drawer. For this paper would eventually find its way to Rome, perhaps through the hands of a zealous student eager to curry favor with the Vatican.

Sometime after his return from Belgium, Teilhard went to southwestern France, accompanied by the Abbé Breuil and the British prehistorian Dorothy Garrod, who several years later would make her great discoveries in the caves of Mount Carmel. They visited many prehistoric caves, climbing into difficult entrances, crawling through narrow passages, excavating for fossils embedded in limestone. This was exhilarating work for Teilhard; since early childhood he enjoyed being in the field. He admired Cro-Magnon cave art and was taken with its beauty. One of the caves the three visited was Niaux, in the Pyrenees, which has one of the most stunning collections of Magdalenian art, from about 15,000 years ago. Back in Paris, Teilhard devoted himself more energetically to his work at the museum, analyzing fossilized remains.

Teilhard knew that because of his status as a scholar, everything he said in public would be monitored by Jesuit authorities in France and in Rome. As his work progressed, he grew even more interested in Darwin's theory and passionate about his own view of evolution. He read other controversial writers on the topic, among them Charles Peirce and Josiah Royce, whose ideas were gaining momentum within the scientific community. These writers influenced his thinking.

As part of his work since the end of the war, Teilhard had been sorting out, cleaning, and identifying the poorly labeled specimens that came to his laboratory at the Institut Catholique from China. The Jesuits had a long presence in China, where they first went as missionaries in the sixteenth century, and had an established infrastructure there, which included schools, places of worship, and museums. One of the museums was at the Jesuit school in Tientsin (Tianjin), on the eastern coast, southeast of Peking. The museum was managed by the priest-explorer Émile Licent, who collected fossils and geological finds. It was Licent who had sent the fossils to Paris.

The workings of the Jesuit superiors in Rome are opaque to an outsider, and the secrecy that governs them has not changed in the three-quarters of a century since Teilhard's actions were discussed there. But the outcomes of their deliberations are known to all. In the early 1920s, Vatican officials were becoming wary of Teilhard's writings and lectures, which preached evolution and rejected conventional notions of Adam and Eve, the Garden of Eden, and original sin. Jesuits found it unacceptable to have one of their priests, who held a position of great visibility in Paris and commanded a growing audience, promulgating views that were anathema to the Catholic Church. These authorities decided to press Teilhard to leave Europe, hoping that his absence from the continent would stop the dissemination of his views.

In an effort to match his interests with the positions the Jesuits had available outside Europe, his superiors suggested that Teilhard move to China and join Father Licent. Teilhard's sister Françoise, who had been a missionary in that country and had recently died, was buried in Shanghai, and the assumption was that Teilhard might be inclined to visit her grave and see the country where she had lived. In Tientsin, Teilhard could continue his work on paleontology with Licent.

Teilhard was not interested. While the Chinese fossils held some professional attraction for him, he was very attached to Paris. He wanted to stay where he was. He felt that his mission was to reconcile the Church with modern scientific theories, and he knew that Paris was the best place for him to pursue this aim. Rome, meanwhile, accused him of daring to propose a "biological philosophy" that was in conflict with the Aristotelian-Thomist doctrine of the Church.

Teilhard felt the mounting pressure from his order to renounce his "heretical" views. It was an impossible situation, and his position in Paris was becoming untenable. As a Jesuit, loyal to his order, he had to make a decision—he could not wait much longer. The Jesuit superior general, Vladimir Ledochowski, was known for taking strong action against priests who did not follow rules, and Teilhard knew that if he remained a Jesuit, obedience was demanded of him.

On October 6, 1922, Teilhard relented. He sent a letter to Émile Licent in Tientsin, saying that he would go to China. But he was committing to stay for only a short time, and he wanted to negotiate the length of his tenure. He also wanted the Institut Catholique in Paris to hold his position for him while he was away; he thought that he could exploit a relatively short stay in China to advance his career, by using his time to study fossils that were being discovered there, and then return to Paris a more accomplished scientist. Teilhard was now backed against the wall, pinned to it by the mighty Church. The Jesuits decided that their recalcitrant priest should stay in China for at least a year. In April 1923, Teilhard sent a telegram to Father Licent in Tientsin: "Come for a year. When should I leave for China?" Licent responded in similar style: "Arrive 15 May."

Licent, who was from the area of Lille in northern France, had an innate coldness, and was not known for his interpersonal skills. But he liked Teilhard enough to have sent him an enthusiastic letter welcoming him to his museum in Tientsin and offering him a stay of anywhere from six months to two years. This fit well with Teilhard's plans.

In a letter to Father Auguste Valensin, a friend from his time in Aix, Teilhard described the hard choices he faced in trying to decide how long to stay away from Paris: "I am forced to choose between two opposing ideas: the one, the rather 'brutal' thought that nothing in life really matters except God; the other, an ever-sharpening awareness of how heavy-handed, narrow-minded, and weak is the modern Church. Sometimes, I find myself thinking, 'I want to be dissolved' in order to escape this inner tearing."

Teilhard didn't dissolve—instead he wrote more papers expounding his belief in evolution and explaining that God works through evolutionary processes to propel humanity ever forward. He was embroiled in an internal conflict between obedience to his church and his own integrity as a scientist and a thinker. In the years to come, this conflict would intensify.

ÉMILE LICENT, who had founded the museum at the Jesuit house in Tientsin, had named it Hoang-Ho Pai-Ho Museum (Museum of the Yellow River and the White River, otherwise known as Huang Ho and Pei Ho, respectively), after the longest and shortest rivers in northern China. Informally, everyone knew it as "Licent's Museum," and for all intents and purposes, Licent ran it as his private property. It bore the marks of his eclectic taste in natural history, and he alone decided what to collect and how to display it. The museum included geological, botanical, mineralogical, and paleontological collections, which he had begun to assemble as early as 1914. He made frequent trips of hundreds of miles each, to Mongolia, Manchuria, and the Tibetan Plateau, to collect fossils and artifacts for study and display.

Once he arrived in China, Teilhard would travel with Licent on these rough collection expeditions throughout Asia. Licent expected his new confrère to be a competent collaborator and a "travel companion who would not prove demanding (more than I am) about food and the comforts of life." Coming from Paris, Teilhard had no realistic idea about what to expect—he could not even guess how spartan his life with Licent would be.

On April 6, 1923, Teilhard boarded a ship in Marseille. The ship sailed through the Suez Canal and into the Red Sea. From the deck, Teilhard saw the Sinai Peninsula with its mountains on his left and imagined Mount Sinai farther to the east. "How I would love to have scrambled ashore," he later wrote his cousin Marguerite, "and tested those rocky slopes. But not only with my geologist's hammer. I would love to have learned if I too could hear the voice of the Burning Bush."

Exiting the Red Sea, the ship traversed the Indian Ocean, stopped in Colombo, passed through the Strait of Malacca between the Malay Peninsula and Sumatra, and continued to Saigon, in what was then French Indochina. Teilhard spent his time aboard ship reading, writing, and observing nature. He liked to look at the stars at night—so clear and bright when seen from a ship far from the intruding lights of terra firma—and by day observe the state of the ocean, calm at times and stormy at others. When the ship arrived in Shanghai, Teilhard disembarked and visited his sister's grave. This was an emotional moment for him, as he knelt by her tomb, next to the red-brick convent to whose mission, tending to China's sick and dying, she had devoted her life. He then went by train to Tientsin—a journey of some four hundred miles.

China was a land in turmoil. The revolutionary leader Sun Yat-sen had been deposed and was kept prisoner on a gunboat off the coast of Canton. The country had disintegrated into separate local communities governed by warlords, who battled one another for dominance.

Pierre Teilhard de Chardin now began a most important—and most trying—period of his life. He had grown up in the genteel world of wealthy French estate owners, studied and become a recognized scientist as well as an ordained Jesuit priest. He had valiantly served his country in war, showing immense courage and determination. In Paris, he had enjoyed the fruits of peace, and the qualities that had made him willing to sacrifice his own life to save soldiers helped him grow and flourish professionally and spiritually. But fate and the response to decisions he had made now sent him to a faraway land torn in conflict. Here he would make a new beginning, and live up to the great scientific challenge life presented him.

Chapter 7

A DISCOVERY IN
INNER MONGOLIA

Teilhard arrived at the Jesuit school of Tientsin and walked into Licent's workshop. The two had met before, in Paris in 1914, when Licent was there to appeal for funding for his operation, but Licent definitely had the upper hand here: Teilhard was entering his territory. Licent was tall, thin, and square-faced, and his white hair was usually unkempt. His black cassock was often covered with dust from work and travel. He was a tireless traveler and adventurer, forever digging for fossils and other relics for his museum. He often embarked on long expeditions to inaccessible locations throughout China, accompanied by guards and coolies, well armed against the highwaymen lurking in the deserts and steppes of Asia. He would gather large numbers of fossils, load them on wagons, and haul them to Tientsin; each trip took months to complete. Though a priest, he always slept with a loaded pistol nearby.

According to the Chinese paleontologist Jia Lanpo, Father Licent was interested in digging for fossils in the deserts of Inner Mongolia because he believed that this region, prehistorically extremely fertile and supporting a great abundance of fauna and flora, was the lost Garden of Eden.

The emperor had given Tientsin the status of treaty port, which meant that foreigners were allowed to live in certain districts of the city, called concessions, in which they enjoyed extraterritorial rights of settlement. And it was on Race Course Road, within the French concession in the city, that Licent had built his museum, attached to the school.

Arriving here, Teilhard suffered from culture shock. He missed Paris, with its universities, theaters, music halls, and cafés, and his many friends. In letters to them he described the wretchedness of the Chinese who lived around the school, who, he believed, had no idealism or hope. But he felt a deep compassion for them, and as he was an exceptionally friendly man who met new people easily, he soon had a number of Chinese friends and began to understand something about their culture. Throughout his time in China, the Chinese would always treat Teilhard with warmth and respect—despite the fact that he never learned their language. His English was good, and he used it for communicating with most people. And French also proved useful in the East.

On June 6, 1923, Teilhard went to Peking to present a paper at a meeting of the China Geological Survey about his analysis of Licent's latest paleontological discoveries. Here he met the people who within a few years would be players in the great scientific adventure to take place in China. These included the Canadian anatomist Davidson Black, the Swedish geologist Johan Gunnar Andersson, and the Chinese anthropologist Pei Wenzhong. Teilhard was pleased to meet them, invigorated by making friends and engaging in conversations about science and the exciting possibilities of paleontological research in China. He would keep his friendships with these Western and Chinese scholars for the rest of his life.

Teilhard returned to Tientsin on June 8. Later that month, he embarked on his first desert expedition. In 1920, Father Licent had uncovered a rich deposit of fossils of a long-extinct three-toed horse in a plateau in eastern Gansu Province. He also discovered, at another location in Gansu, three worked-stone tools from the Upper Pleistocene (the latest geological period within the Pleistocene era, stretching from 125,000 to 11,500 years ago), in a loess deposit, a yellowish-brown foamy layer of dust left by the wind and then hardened. These were the first stone implements to be discovered in China, and the find was of great significance because it ended the false view in archaeology that China contained no Stone Age human industry and supported no Stone Age humans. Licent's success spurred him to search for fossils that would indicate human presence.

In 1922, Licent traveled to Mongolia after he heard a rumor that bones had been found in the Ordos Plateau, near the Yellow River four hundred miles west of Tientsin. Searching the deserts of Inner Mongolia, Licent discovered the important site of Sjara-Osso-Gol (Salawusu), located at the southeastern corner of the plateau, bordered by a southward bend of the Yellow River. In sand deposits dated to the Pleistocene, he found a large number of fossil animal bones. This discovery made him decide that the site deserved a more thorough exploration. And now that he had the help of an able paleontologist like Teilhard, it was time to look for human fossils.

In mid-June 1923, Licent and Teilhard took the train to Hohhot ("The Blue City"), in the steppes of Mongolia, where they purchased letters of passage and picked up mules and supplies, servants, and a military escort. They then headed into the Ordos Desert toward Sjara-Osso-Gol.

It was soon apparent that this goal would be difficult to achieve. A local warlord had amassed a force of six hundred soldiers of fortune and was on a hunt for "foreign devils," whose heads he vowed to place on poles throughout the Ordos Desert. So Licent and Teilhard turned their caravan north, heading deep into Mongolia. As they penetrated farther and farther north, the terrain was more arid and barren; but every once in a while they found themselves near an oasis with trees and a lamasery inhabited by Buddhist monks dressed in saffron-colored robes. At times they were stalked by Mongolian horsemen, who stared at them with hostility but kept their distance, since it was evident that the Jesuits and their guards were heavily armed.

The group continued west along the north bank of the Yellow River, and turned south, crossed the river, and proceeded along its western bank. At Hengcheng, southeast of Yinchuan, the capital of the Ningxia Hui Autonomous Region, they camped for several days and surveyed the terrain in search of fossil-bearing formations. Then they went east, traveling parallel to the Great Wall. Fifteen miles east of Hengcheng, in view of the Wall, Licent and Teilhard discovered the Shuidonggou site.

This now famous Paleolithic site is rich in Stone Age artifacts and has been used, since its discovery by the two Jesuits, for comparison of Eastern and Western Stone Age cultures. The tools found here, examples of the stonework technology identified as the Shuidonggou industry, are very similar to the Mousterian and Aurignacian stone implements discovered in Europe. (The Aurignacian industry was a stone-tool-producing culture in Europe, from 36,500 years ago to 28,000 years ago; it was preceded by the Mousterian, which, again, lasted from 250,000 to 40,000 years ago. The makers of Mousterian stone tools are believed to have been mostly Neanderthals.)

The Shuidonggou site has been excavated four times after its original exploration by Licent and Teilhard, most recently by a team led by Gao Xing of the Institute for Vertebrate Paleontology and Paleoanthropology in Beijing, and has yielded thousands of stone tools, among them some microliths—very small blades of flaked stone—and ground stones. Along with the stone tools, teeth and fragments of fossil mammals have been found. The site has been dated with radiocarbon analysis, yielding an estimate of about 30,000 years ago; it is concurrent with the Cro-Magnons and the last days of the Neanderthals in Europe. The original site discovered by Teilhard and Licent is now seen as part of a much larger Stone Age area of Inner Mongolia, which includes locations that have yielded remnants of Upper Pleistocene human habitation.

In their 1923 expedition, the two Jesuits discovered a vast number of stone tools, which Teilhard later described in letters to friends and colleagues in Paris. At Shuidonggou, Licent and Teilhard lodged in a tiny inn with two small, sparsely furnished rooms for travelers, the innkeeper and his family cramped together into a third room. The meals the innkeeper's wife prepared for them daily were meager, food being scarce in the region. It was therefore a great treat for the two men when she was able to offer them a meal of eggs and potatoes.

After unearthing a large number of animal fossils and stone tools, the Jesuits and their party continued to Sjara-Osso-Gol. The Mongolian name of the nearby river, Salawusu, means "yellow water," and indeed, the river that flows here, a tributary of the Yellow River, is always filled with silt, which makes the water appear yellow.

Never before had Teilhard felt so in tune with nature's rhythms, and this experience was especially moving. It was in the Ordos Desert that he wrote his "Mass on the World." This beautiful prayer begins:

> Since once again, Lord—though this time not in the forests of the Aisne but in the steppes of Asia—I have neither bread, nor wine, nor altar, I will raise myself beyond these symbols, up to the pure majesty of the real self; I, your priest, will make the whole earth my altar and on it will offer you all the labors and sufferings of the world.

Religion, science, and nature were inexorably intertwined in Teilhard, creating a personal mysticism. From the desert, he wrote to the Abbé Breuil: "Mysticism remains the great science and the great art, the only power capable of synthesizing the riches accumulated by other forms of human activity."

Teilhard described nature and what he saw and did in the Ordos to his friends in Paris: "We're surrounded by horses, kites, and cranes, as tame, almost, as garden pets. It's altogether bucolic. The Mongols wear long hair, never take off their boots, are never out of the saddle, and dislike cultivating the soil. The Mongol women look you straight in the eyes with a slightly scornful air, and ride like the men. . . . The digging goes on. I have formed an exact idea of the geological formation in which we are (it is fairly recent for China), and I attach a great deal of importance to these conclusions. We live under canvas, dressed only in shirt, trousers, and Chinese jacket. It's the real free life!"

The two Jesuits found the remains of thirty-three species of mammals and eleven species of birds. These included fossils of the straight-tusked elephant, woolly rhinoceros, Ordos giant deer, giant ostrich, antelope, and other Upper Pleistocene Chinese fauna. One of the species they named Wansjock's buffalo, in honor of a local Mongolian whom Licent had met years earlier and whose information had first led him to Sjara-Osso-Gol.

Here, Licent and Teilhard lived in a tent, and when it rained, Teilhard stayed inside, writing. He described the discoveries the pair had made earlier at Shuidonggou. That site, more than this geologically older one, held the greatest promise of finding human remains—there was no escaping them, given that an extensive stone-tool industry was in evidence there.

In his letters, Teilhard described how he and Licent had unearthed cooked gazelle bones, fossils of woolly rhinoceroses and giant ostriches, and chipped flints similar to the Cro-Magnon stone tools Teilhard had seen in Europe. They found giant deer antlers that had been broken for use as tools, and an extensive set of quartzite cutting blades. Teilhard was euphoric about the discovery of signs of early human habitation in such a remote and desolate corner of Asia. On August 15, 1923, he wrote a friend, the Abbé Christophe Gaudefroy, in France:

I am writing you under the tent, on the bank of a curious little river that runs at the bottom of an 80-meter-deep canyon, in a region of steppes and dunes. We have arrived here (nominal destination of our

trip) only ten days ago . . . the combined action of drought and brig-
ands having forced us to divert our route toward the northern tributary of the Hoang-ho [Yellow River] before making a straight pass
through to the point at which we are now (see map at the end of the
letter). We had gone by train (what a train!) until the terminus (Pao-
to [Baotou]), and there mounted a caravan of ten mules, with which
we have been traveling since the 22nd of June, half as mandarins, half
as soldiers—dressed in khaki, with many rifles. I am not upset about
this long journey in Mongolia, which has shown me much terrain, and
made me find many things I had not expected: for example, mammals
of the Pliocene [a geological era from 5.3 to 1.6 million years ago]
(until recently unknown in China) and a Paleolithic hearth (something equally new in China); about that last point, I wrote to Breuil
two weeks ago. At present we are discovering numerous bones (rhinoceros, horses, ox, gazelles, camels, hyenas, wolves, etc.). . . . It's a
lot of work, and we employ twenty Mongols and Chinese.

Indeed, Teilhard had written to the Abbé Breuil a month before, on
July 16, 1923, from the camp at Sjara-Osso-Gol:

Very Dear Friend,

I haven't given you a sign of life, despite my promise, since my
card of June 12 from Tientsin.—It is because travel leaves much less
time for leisure than I had supposed. Today, the rain is keeping us
inside the tent. I am using the opportunity to write to you, at least
as much to satisfy myself as to give you satisfaction. I've never felt so
well as during my isolation. . . . I write from point A on the rough
map here enclosed. All kinds of difficulties, climatic (drought that has
stopped the growth of desert plants) and political (the presence of
numerous bandits at the north bend of the river) have in effect forced
us to modify our route. . . . I have made . . . three interesting studies:
two of pure geology on the mountains north of Pao-to and on the

Arbous-oula [Zhuozi Shan]. . . . This deposit of Chara-Ousso [Sjara-Osso-Gol] is much more important to untangle since Licent has collected here a human femur (quite special) as fossilized as the rhinoceros, and a fragment of bone clearly worked on.

In his next letter to Breuil, Teilhard related discoveries he had made of remains of ancient animals including bison, mammoth, and hyenas, and continued with great excitement: "And man?—He is *surely* here. But here I must proceed methodically." He described signs of ancient human stone industry in the region. Teilhard made more precise scientific observations about early human tools produced here during the Paleolithic era in a letter to his scientific mentor at the Museum of Natural History in Paris. As he told Marcellin Boule, after returning to Tientsin:

For over ten days we have made the most of the famous hearth discovered at the end of July [at] Ning-Hia-Fou [Yinchuan]. Thanks to the habitual ingenuity of Father Licent, we have been able to exploit 80 square meters of surface (to a depth of 10 meters). Result: more than 300 kilograms of cut stones, literally strewn over the habitation area. At least a hundred of the instruments are admirable (in their fashion), some of them are of a "colossal" type: scrapers/hull-shaped slats, 10 centimeters long with large triangular fistlike handles; shear-scrapers as long and as wide as my extended hand; square scrapers as large as fists. There are some admirable "Mousterian" points.

Thus Teilhard and Licent discovered in China a stone-tool industry that resembled the Middle Paleolithic industry familiar in Europe. This was an important development, and it gave Teilhard the experience he needed for the crucial scientific role he would play in the discovery and analysis of Peking Man.

The fossils that he and Licent had found at Shuidonggou were in

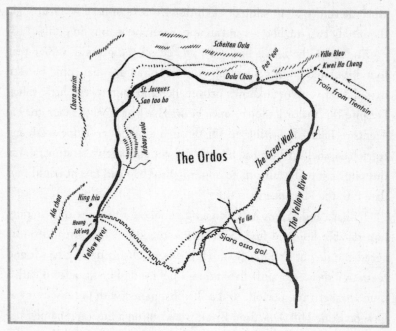

Teilhard's route in the Ordos, Inner Mongolia. The map includes place names as they appear in Teilhard's writings: "Ville Bleu" is Hohhot, Peo t'eou is Baotou, Arbous oula is the Zhuozi Shan range, Ning hia is Yinchuan, Sjara osso gol is Salawusu. Map reprinted from Thomas M. King, S.J., *Teilhard's Mass: Approaches to "The Mass on the World."* Used by permission.

nearly perfect condition and included whole skulls and complete skeletons of rhinoceros and buffalo. In addition, the stone tools, and the horns that seemed to have been cut and used as tools by early humans, provided strong evidence for a stone industry in this location. At Sjara-Osso-Gol, the pair found scrapers and points of similar workmanship, as well as animal remains—some intact, and others that obviously had been butchered by early humans.

These two sites together proved scientifically invaluable in establishing the existence of an ancient human presence in China. As Teilhard's letters indicate, however, he and Licent were not content to have made these discoveries. They wanted the big prize: human skulls—the ultimate proof that humans had once lived here. They were therefore disappointed when those weeks of hard work brought them only two thighbones and an upper arm bone, but no skulls.

And since the fossils had been picked off the ground, rather than found within a given geological sediment of an age that could be estimated, they were of unknown provenance and therefore of little value. Despite this major letdown, once he was back at Licent's laboratory in Tientsin, Teilhard was able to sift through a pile of antelope teeth and ostrich eggshells, which had been found within a known stratum, and in this pile he found a humanlike upper incisor. He and Licent could thus date it to the Pleistocene.

Teilhard and Licent now had a humanlike tooth—the only important datable hominid find from their expedition. This situation mirrored the one at Dragon Bone Hill, where Otto Zdansky had found teeth to indicate hominid habitation—believed to be associated with a much earlier time period. And as had happened with the discovery at Dragon Bone Hill, Davidson Black went out on a limb and named the unknown hominid whose tooth Teilhard and Licent had discovered "Ordos Man." Further relics of Ordos Man would be discovered in the 1970s and 1980s by teams working under the sponsorship of the Chinese Academy of Sciences—making Teilhard and Licent's work in the 1920s even more significant. But modern dating techniques now place these finds in the Upper rather than the Middle Paleolithic, as the original discoverers had posited (meaning that they are less than 45,000 years old).

These profound discoveries propelled Teilhard, already an accomplished paleontologist and geologist after his work with Boule, into the front line of paleontological research. He was now an accomplished field researcher as well as a scholar—one who could identify likely sites, date them with some accuracy, and make significant discoveries. The best fossil hunters today, scientists who have been uncovering human origins in Africa, Asia, and elsewhere, possess a similar combination of skills.

Teilhard and Licent floated down the Yellow River on a barge, covering quickly the distance that had taken them weeks to traverse on land. They returned from their adventure in late September, more than three months after they left Tientsin. Teilhard was eager to return to France and incorporate his findings into new research at the Institut Catholique and the museum. He was looking forward to a reunion with his *maître et ami*—"master and friend"—Boule, and to sharing with him the striking discoveries that he and Licent had made.

Teilhard wrote to his provincial in Lyon, requesting permission to return to Europe. The response from France was that the order preferred for Teilhard to stay longer in China. Deeply disappointed, he fell into a depression. Yet if the Jesuit authorities thought that by keeping Teilhard in China they would cool his zeal for the study of evolution, they were to be disappointed, for his interest only grew stronger. The tantalizing discoveries of signs of early human habitation in the Ordos Desert whetted his appetite for finds with deeper evolutionary relevance.

In China, far from the eyes of petty Church autocrats, he was free to pursue his paleontological interests in search of human origins. Teilhard was now familiar with the work of his new friend Andersson at Dragon Bone Hill and the discovery there of humanlike teeth—so similar to his own discovery in the Ordos Desert. Both Andersson and Davidson Black showed heightened respect for the priest-scientist and were eager to include him in their international group of experts now poised to make one of the greatest discoveries in twentieth-century anthropology.

Teilhard braced himself for a hard winter in Tientsin. February brought heavy dust storms with sand that hurt his eyes. He found solace in his work, putting in many long hours at the museum, classifying and analyzing the fossils collected on the Mongolian expedition and sending reports to the museum and the Institut Catholique in Paris.

Summer finally brought good news: Teilhard would be allowed to return to Paris. The measure of his relief and happiness at the prospect of going back can be glimpsed in lines he wrote to a friend there: "I, too, often dream that I am back in my customary place in your charming dovecote! . . . This is probably the last letter I will write you from China." Teilhard was elated and full of hope on September 13, 1924, as he boarded a ship in Shanghai, after praying at the tomb of his sister. A month on the ocean allowed him to clear his head, write down his thoughts about religion, and detail and further summarize his paleontological discoveries in the Ordos Desert. Relaxing on a deck chair, he spent many hours looking at the sea and thinking. As the journey ended, he was refreshed, ready to return to his old lifestyle in the West. He landed in Marseille on October 15, and took the first train to Paris.

His seventeen-month sojourn in China had given him much: experience in paleontology and knowledge of the terrain, from the Pacific coast to the deserts of Mongolia. But he felt that he belonged in Paris. He was excited about restarting his life in the French capital, putting to use his new knowledge and data gathered in China, and resuming contacts with his many friends, students, colleagues, and admirers. Friends flocked to see him at his room in the Jesuit residence and at his office at the Museum of Natural History, where Boule, pleased to have Teilhard back, wasted no time in assigning him the task of supervising the analysis of the many fossils he had sent from China, including those of extinct rhinoceros and hippopotamus species. Teilhard oversaw the work on these fossils by two doctoral candidates in paleontology.

Teilhard also resumed his lectures, and found that the enthusiasm for his ideas had only intensified since he had been gone. He had a loyal following of students and colleagues; to young Jesuits and others, he was a leader whose views were novel and progressive—a breath of fresh air within a stagnant religious establishment. His new approach to religion was appealing, as it incorporated science and other aspects of contemporary life within a Christian context.

Many people in Europe thought that Teilhard was the leading

thinker the Jesuits needed—one who could move the Church forward and recast its principles in a vibrant, modern setting. Within the world scientific community, he had a first-rate reputation, now bolstered by the discoveries he had made with Licent in Mongolia. Equally, he was highly respected and revered by many within the Church. Young Jesuits who had heard him discuss reconciling the carnal world with Christian morality felt that "Father Teilhard speaks, and it is immediately evident that he counts among the intrepid ones, not the innocents."

During this period of intellectual excitement and creativity, Teilhard again recorded his ideas on evolution. The document he wrote set forth his overarching scheme of the universe, a giant entity evolving from rocks to life forms to complex organisms to intelligent beings, and on to a living, thinking planet. This idea led Teilhard to conceive the noosphere, which he imagined as encompassing Earth, covering its biosphere—the sphere in which living things exist—surpassing it, and containing the essence of intelligence and thought.

It was an elegant idea, one that combined science with spirituality. Teilhard's understanding of biological evolution was scientifically correct; he knew how the forces of evolution work on living systems. And he was now openly evolutionist in his views, despite Church opposition to Darwin's theory. Teilhard expanded this science into the realm of spirituality and mysticism, positing below the scientific structure of biological evolution a similar, perhaps metaphoric, force affecting inanimate objects as well. And above this structure, he constructed a divine edifice that used evolution as a base for propelling both human intellect—the ultimate product of biological evolution—and our planet as a whole. In his own way, Teilhard wedded science with faith.

Teilhard's troubles with the Church had not gone away—they moved to a higher level. While he was formally allowed to publish on science, his new writings on evolution, which appeared in the journal *Revue de Philosophie* in late 1923, exacerbated the situation. His Jesuit superiors in Rome had by now read the paper he had written after his visit to Belgium a few years earlier, "A Note on Certain Historical Representations of Original Sin"—which is now believed to have been stolen from his drawer at the Institut Catholique—and they were furious. We don't know exactly what was in Teilhard's paper, just that it challenged the orthodox view of original sin. The paper was read in the Vatican by none other than the secretary of the Holy Office, Cardinal Rafael Merry del Val, who at once sensed heresy and approached the Jesuit superior general, Vladimir Ledochowski.

On November 13, 1924, only weeks after his return to his homeland, Teilhard received an urgent letter from his provincial, Father Olivier Costa de Beauregard, summoning him to Lyon. A messenger had arrived there from Rome, carrying papers that Ledochowski wanted Teilhard to sign. These documents contained six separate propositions. All we know from Teilhard's letters and conversations with friends (prime among them Father Auguste Valensin) is that he was ordered never to say or write anything against original sin, and that he found the fourth proposition—whatever it was—the most difficult to accept.

These demands precipitated a severe crisis for Teilhard. Because he was now such a prominent scientist and philosopher, he felt he could not be a leader in his fields if he was prevented from being a free-thinker. "They want me to promise in writing," he wrote Valensin, "that I shall never *say* or write anything against the traditional position of the Church on original sin. This is at once too vague and too absolute. . . . In all conscience, I believe I must reserve (1) the right to search, and (2) the right of support. I hope to be able to reach a way of commuting the formula they want me to sign."

Teilhard's dilemma was how to reconcile the open, free mind of a scientist and philosopher with the blind, total obedience demanded by the Jesuit order. His friends suggested that he fight the demand; some even counseled that he leave the order and pursue an independent career as a scientist. After all, they argued, he already held an important position as both scientist and philosopher: he could continue his work without being a member of the Society of Jesus.

But Teilhard knew that priests who left the order generally did not fare well in society and their careers often faltered. Moreover, he was not raised to take such a step. He was from a tradition that held that when a person gave his word, he must keep it. In addition, he simply did not believe that he should leave his chosen path in life. He was a Jesuit, a believer, and he had no intention of leaving the order. As he would later write his friend Édouard Le Roy, "I still have just one choice: to be a perfect religious or to be excommunicate." Jesuits follow orders as soldiers do, and Teilhard had made his choice: He would remain a loyal priest. But his life was now unpleasant as well as extremely complicated.

Teilhard took a break from his troubles in Paris, and in April 1925 traveled to England with the Abbé Breuil. The two joined Dorothy Garrod in Suffolk, near Ipswich on the east coast, and explored a steep crag in which they found flint blades indicating human activity that they estimated to have taken place 20,000 years before. Teilhard compared their findings with those he had made with Licent in China. On the way back to Paris, Teilhard and Breuil spent a few weeks in Sarlat-la-Canéda in the Périgord, exploring the caves of Les Combarelles and Font-de-Gaume, at which multicolored cave art of horses, bison, and deer had been discovered around the turn of the century. Breuil had been associated with the discovery, and had authenticated the dating of the art to the last Ice Age.

In May, Teilhard wrote to Valensin, complaining about his dire circumstances vis-à-vis the Church. His order was now pressing him to leave Paris, so that he and his unorthodox views would receive less coverage and less attention. If he lived in some province, away from the limelight, he would be much more tolerable to Jesuit authorities.

His resistance was wearing down from the constant pressure, and he understood that he had to act quickly to end this impossible situation.

At some point in 1925, he signed the six propositions demanded by the Jesuits. Teilhard scholars today, Jesuits included, do not know the exact content of these secret documents. Eighty years after the event, the papers Teilhard de Chardin signed are still kept locked up in Rome.

Teilhard confided to Valensin that the papers he had been forced to sign, including a promise never to say anything contrary to the Church's traditional view on original sin, were apparently not enough to appease his superiors. They still distrusted what they saw as a renegade priest, and they remained determined to keep him away from where they felt he could do them harm. Teilhard began to wonder whether the only way to be left alone was to return to China.

Chapter 8

AUSTRALOPITHECUS
AND THE SCOPES TRIAL

D arwin believed that Africa was the cradle of the human lineage. But until early in the twentieth century, nothing had been found on that continent to merit scientific attention. In 1913, the German paleontologist Hans Reck was working in what is today Tanzania, at that time called German East Africa. He was looking for bones in a wide gorge known as Oldoway. (The name would later be altered to Olduvai.) There, among fossils of extinct animals, Reck discovered a skeleton that seemed to belong to a human. But scientists informed about the find dismissed it as the remains of a buried modern human that had simply intruded into the site of ancient animal bones. The discovery did not receive the serious notice it deserved as hominid remains until after Louis Leakey made his own discoveries in Olduvai Gorge in the 1930s and after.

In June 1921, another notable fossil was found in Africa. Zinc and lead were being mined at Broken Hill in Northern Rhodesia—now Zambia. A mine worker, T. Zwingelaar, was digging sixty feet belowground when he came upon a skull. It looked like a human skull, but it had massive brow ridges, a low forehead, and an elongated cranial vault—in fact, it was similar to the Neanderthal skulls discovered in Europe. The teeth were intact, and they looked human. They even had cavities.

An engineer at the mine traveled to London a few months later and took the skull to the Museum of Natural History. There the species represented by this fossil was named *Homo rhodesiensis* (Rhodesian Man), and scientists considered it a form of Neanderthal. It generated interest in the scientific community and was inspected by Java Man's discoverer, Eugène Dubois, who was intrigued by the possible relationship between his fossil and the new find. But the enthusiasm for the discovery remained muted, even within scientific circles. And there was confusion about some bones that at first were said to have been found with the skull but later were determined to have come from a different source and arbitrarily lumped with it. Some of the fossils appeared more archaic than those of Neanderthal, while others looked modern.

In 1924, while Johan Andersson was finalizing his plans for expanded excavation at Dragon Bone Hill in China, Raymond Dart, an Australian anatomist and anthropologist who was teaching at the medical school of the University of the Witwatersrand in Johannesburg, was alerted to the discovery of an interesting fossil. Dart, who had been teaching at the medical school for more than a year, had set up a project of comparing skeletons. He asked his students to look for fossils to bring to class for analysis. The only female student in the class, Josephine Salmons, mentioned that miners at the Buxton Limeworks quarry near Taung, in the northwestern region of South Africa, had discovered an unusual fossil. She had visited the home of the quarry's owner, and on his desk had seen what she took to be the skull of a baboon.

Dart made arrangements with the quarry management to send him a box of normal rubble, so that he could sift through it. Soon after he started looking through the debris he had received, he found a skull that looked human. It was small, however, and he concluded that it had belonged to an immature individual—a child.

Dart worked for more than two months to remove the rocky matrix—the substance enclosing the skull—using a hammer and one of his wife's knitting needles. When he finished, he had before him the well-preserved skull of a young hominid—with a permanent first molar just emerging. If this had been a modern child, the age would have been about four or five.

Dart immediately understood the great importance of his finding: The young creature was a member of a yet unknown species that developmentally was between apes and humans. He realized that the cranial capacity was greater than that of an immature ape, but much smaller than that of a modern human child. He also noticed that the milk canines were smaller than those of chimpanzees. This indicated that the species had a different diet from that of chimpanzees—its teeth were evolving to become more like human teeth.

The child's foramen magnum—the hole at the bottom of the skull, through which the spinal cord connects to the brain—was at a more forward location than that of apes. This, too, was a humanlike sign, since the human is the only animal to walk upright and have a head that is balanced directly on top of the spinal column. The child's skull indicated a species that walked upright at least part of the time.

Dart believed that he had discovered the missing link between humans and apes. He named this new species *Australopithecus africanus* (African Southern Ape), but stressed that this "ape" had many human-like features, such as the small teeth, the larger cranial capacity than an ape, and the more upward posture as inferred from the location of the foramen magnum. Dart's paper about this find appeared in the February 7, 1925, issue of *Nature*. Later, the term "australopithecine" would be used to refer to *Australopithecus africanus* and several related species, whose fossils are still discovered today.

Dart's specimen was later dated to about 2 million years ago, and was thus the oldest hominid discovered so far. In the following decades, other kinds of australopithecines would be discovered, some of them older than Dart's fossil, and all of them representing an important step in the evolution of humans.

While some in the scientific community embraced Dart's findings, others did not. In the issue of *Nature* that followed his article, several scientists wrote that he had mistaken an immature chimpanzee-like creature for a human ancestor. Although they had not seen his find, they accused him of dating the skull incorrectly; it might have come from a more recent geological level. Dart was "complimented" for discovering a new species of ape, rather than a creature closer to humans; he had, his critics suggested, recovered bones of recently deceased animals that had fallen into a cave, and had mistaken them for ancient relics. Dart vehemently denied these charges and forcefully argued that his find was authentic, comprising the integral remains of an animal that was neither human nor ape but something in between.

At that time, most scientists believed that human ancestors might be found in Asia, not in Africa, and this notion convinced many that Dart was wrong. Naysayers dismissed his discovery, nicknamed "Taung Baby," as the skull of a chimpanzee or a baby gorilla. The world was not ready for it, and so Dart's fossil and the earlier African discoveries were soon forgotten.

Years later, Dart would open the jaws of the Taung skull, revealing all its teeth, and new analyses would support his original claim. And when over the following decades other fossils were discovered belonging to the same species, Dart and his views would be vindicated.

One scientist who believed in what Dart had claimed was Robert Broom, an eccentric Scottish paleontologist and physician who had conducted important research on lizards, dinosaurs, and early mammals. Because of this work, he was elected a fellow of the Royal Society in 1920. Two weeks after Dart's article about *Australopithecus* appeared in *Nature*, Broom arrived in South Africa, burst into Dart's lab, and knelt in front of the Taung Baby "in adoration of our ancestor."

A decade later, Broom gave up his medical practice to conduct paleontological research in Africa full-time. He found other australopithecines in the South African caves of Sterkfontein, which Teilhard de Chardin would visit. In a cave near the farm of Kromdraai, Broom discovered another australopithecine, *Australopithecus robustus* (which he originally named *Paranthropus robustus*, or Robust Near-Man). The *robustus*, or robust australopithecine, had molars that were much larger than those of *africanus*, and the cheekbones were more widely flared, indicating bigger facial muscles. *Australopithecus africanus* lived roughly 3 million years ago, and *Australopithecus robustus* lived between 2 million and 1 million years ago.

Both species were small-bodied, and their cranial capacities were not much greater than those of chimpanzees—about 500 cc for the australopithecines, compared with a chimpanzee's 400 cc. Theirs was about a third of human cranial capacity. Even for their small size, the australopithecine brain was small.

Later in the twentieth century, older australopithecines would be discovered, known by such names as *Australopithecus afarensis* (Southern Ape of Afar) and *Australopithecus anamensis* (the species name refers to the word for "lake" in a local African language). These would be dated from 4.2 to 3 million years ago, placing them further back in the evolutionary line toward the divergence of our lineage from that of the apes. The *Australopithecus afarensis* fossils include the famous "Lucy" discovered by Donald Johanson in 1974, and "Lucy's Baby," a 3.3 million-year-old fossil (older than her 3.2 million-year-old "mother") of a three-year-old girl whose discovery was announced in 2006. The earliest hominids known today are close to 7 million years old.

Australopithecus is thus a crucial step in the evolutionary succession from apelike creatures to humans. The australopithecines and Dubois's *Pithecanthropus* are missing links—they are placed at various points along our line of evolution. Unfortunately, at the time of Dart's great discovery of the first *Australopithecus*, his breakthrough in understanding human evolution was mostly ignored. One reason for this was that the world needed to get excited about the debate surrounding evolution before it paid close attention to such discoveries. Within a short time, this would happen.

THOUSANDS OF MILES FROM AFRICA, on another continent, an important event in the debate on evolution was beginning. A few months after Dart announced his discovery, John T. Scopes, a twenty-four-year-old football coach and substitute science teacher in Dayton, Tennessee, was indicted for teaching the scientific theory of evolution—a violation of state law.

In March 1925, the State of Tennessee had passed the following act:

Chapter 27, House Bill 185 (By Mr. Butler)
     Public Acts of Tennessee for 1925
     An Act prohibiting the teaching of Evolution Theory in all the Universities, Normals and all other public schools of Tennessee, which are supported in whole or in part by the public school funds of the State, and to provide penalties for the violations thereof.
     Section 1. Be it enacted by the General Assembly of the State of Tennessee, That it shall be unlawful for any teacher in any of the Universities, Normals and all other public schools of the State which are supported in whole or in part by the public school funds of the State, to teach any theory that denies the story of the Divine Creation of man as taught in the Bible, and to teach instead that man has descended from a lower order of animals.
     Section 2. Be it further enacted, That any teacher found guilty of the violation of this Act, shall be guilty of a misdemeanor and upon conviction, shall be fined not less than One Hundred ($100.00) Dollars nor more than Five Hundred ($500.00) Dollars for each offense.

AUSTRALOPITHECUS AND THE SCOPES TRIAL • III

Section 3. Be it enacted, That this Act takes effect from and after its passage, the public welfare requiring it.

Passed March 13, 1925
W. F. Barry, Speaker of the House of Representatives
L. D. Hill, Speaker of the Senate
Approved March 21, 1925
Austin Peay, Governor

John Thomas Scopes, who was born in 1900, studied Darwin's theory in high school and at the University of Kentucky, where it was taught as part of contemporary biology. He also took law courses in college, and graduated in 1924 with a bachelor's degree in law. He was then offered a position as teacher and coach in the school district including the town of Dayton, in the Tennessee River Valley.

Scopes was a quiet, easygoing young man with red hair and a boyish smile. In April 1925, a month after the Butler Act was passed, Scopes was asked to substitute for the regular biology teacher, who was on sick leave for two weeks. Scopes used the textbook assigned to the class, George Hunter's *A Civic Biology*, which had been part of the Dayton school curriculum since 1919. This book explained Darwin's theory of evolution in a matter-of-fact way, couching it within the study of biology. The substitute, who was accustomed to the study of evolution as part of science, saw nothing unusual about this textbook. And while some people may have been vaguely aware of the passing of the Butler Act, there was not much discussion of it; many teachers didn't realize that they were expected to change their curriculum in accordance with the new law.

In the meantime, Lucille Milner, secretary of the American Civil Liberties Union in New York, happened to see a small news item about Tennessee's governor signing the Butler Act into law. She brought this to the attention of the ACLU director, Roger Baldwin, who, with the approval of his board, began to raise money to support a test of the Tennessee law. Newspapers across the country then reported the ACLU's plans.

A manager at the Cumberland Coal and Iron Company in Dayton, George Rappleyea, read a newspaper account of the ACLU's proposed action, and decided that it would be in his and his company's interest to put the small town on the map by focusing national (and, as it would turn out, international) attention on Dayton as the center of a major conflict certain to erupt once the ACLU got its machinery going. The town had been steadily losing population over the years, and its present count of 1,800 was almost half of what it had been around the turn of the century. Media attention could only help, Rappleyea and other citizens felt. Rappleyea contacted F. E. Robinson, president of the Dayton school board, and the two men decided that John Scopes, the substitute teacher using Hunter's textbook, would be the best candidate to test the new law. Scopes was young and without family, and any potential repercussions of the case probably would not cause permanent damage to his career.

A number of leading citizens joined the discussion, and all agreed on the plan to use Scopes to focus attention on Dayton. Scopes was summoned to meet Rappleyea, Robinson, and others, and was asked whether he would agree to be put on trial for teaching evolution. He replied that he could not see how biology could be taught without teaching evolution. And he said that he had already broken the new law by assigning a reading from the biology textbook that included a discussion of human evolution—but coincidentally, he was out sick when the discussion of the reading assignment was to take place, on April 24. But in classes he had taught, he had touched on the topic of evolution. Scopes agreed to be the guinea pig. He later explained his reasons:

Why had I volunteered to be prosecuted for teaching evolution in a public school, thereby violating the criminal code of Tennessee? . . . The answer is heredity and environment. . . . The cause defended at Dayton is a continuing one that has existed throughout man's brief history and will continue as long as man is here. It is the cause of freedom for which each man must do what he can. I did little more than sit, proxylike, in freedom's chair that hot, unforgettable summer—no great feat, despite the notoriety it has brought me. My role was a passive one that developed out of my willingness to test what I considered a bad law. I felt that others did more than I; if men like Clarence Darrow [his attorney] had not come to my aid and had not dramatized the case to a responsive world, freedom would have lost.

Once the ACLU, contacted by Rappleyea and Robinson, agreed to have the test case in Dayton, they invited Scopes to New York to discuss his willingness to be charged in a criminal case. He was offered the choice of a lawyer to represent him (by then he had retained a local attorney, but he accepted the ACLU's offer), and he chose Clarence Darrow, the most famous defense lawyer in America, a man who had just saved convicted murderers Nathan Leopold and Richard Loeb from a death sentence in Chicago. Darrow, who had been approached earlier by the ACLU, had already agreed that if he was chosen, he would defend John Scopes without a fee.

When all the arrangements had been made, F. E. Robinson called a reporter in Chattanooga and informed him that a teacher in the Dayton school, John T. Scopes, had violated the Butler Act by teaching evolution. Thus began one of the most famous court cases in history. Scopes was indicted on May 25, 1925; he was nominally arrested but was never detained. His trial began on July 10, when Judge John T. Raulston called the Rhea County court to order and invited the Reverend Cartwright to lead an opening prayer.

More than six decades after the publication of *On the Origin of Species* in England, Darwin's theory was put on trial in the New World. And this trial, which tested whether society should accept evolution as a theory worthy of being taught to children, showcased what many people saw as an inherent conflict between faith and science. Or rather, it was a test of whether a branch of science, now more than half a century old and with physical evidence to support it, could stand up against a literal interpretation of the Bible.

After the opening prayer, Judge Raulston stated that Scopes was accused of "violating what is generally known as the anti-evolution statute." He then read aloud the Butler Act, which forbade teaching at universities and schools in Tennessee "any theory that denies the story of Divine Creation of man as taught in the Bible." He also read aloud the entire first chapter of Genesis. After some time, he reached the key passage:

> And God said, let us make man in our image, after our likeness: And let them have dominion over the fish of the sea and over the fowl of the air, and over the cattle, and over all the earth, and over every creeping thing that creepeth upon the earth. . . . So God created man in His own image, in the image of God, created He him; male and female created He them.

The jury was charged with determining whether, by teaching evolution, John Scopes had broken the law because the theory he taught contradicted the biblical story of the creation of man and woman in Genesis.

The original prosecutors were Scopes's own friends Herbert E. Hicks and Sue K. Hicks, two local attorneys. Later the prosecution was led by William Jennings Bryan, a three-time Democratic candidate for president and a populist Christian fundamentalist who was on a crusade to ban the teaching of evolution in the United States.

Bryan and Darrow were well matched for a courtroom confrontation. Bryan represented religion and traditional values, while Darrow stood for science and progress. At stake in this trial was not only the central conflict between science and faith, but also the separation of church and state, as well as the very idea of intellectual and educational freedom. The strict interpretation of the passages from Genesis was to be tested against science, and Darrow had lined up a number of scientists to testify for the defense. As it turned out, however, Judge Raulston barred most of them from taking the stand.

At the time, evolution was such a controversial theory that some scientists were not convinced. Opponents of evolution often associated it with atheism. It didn't help the defense that Darrow was an avowed agnostic, while Bryan was perceived as a good Christian and was the author of *In His Image*, in which he argued that evolution was both irrational and immoral.

The trial promised to be a great public spectacle: Every day the courtroom was filled beyond capacity, more than a thousand people attending, three hundred of them standing. This was the first trial in U.S. history to air on national radio, and it drew worldwide media attention. Hundreds of journalists descended on Dayton, to the delight of local businesspeople. H. L. Mencken, who covered the Scopes trial for the Baltimore *Evening Sun*, labeled it "The Monkey Trial."

Henry Louis Mencken, an icon of the American literary scene of the 1920s, was born in Baltimore and studied engineering at the city's Polytechnic Institute before turning to journalism in 1899. Known for his caustic criticism of American culture, he became America's favorite pundit, variously called "The Sage of Baltimore" and "The American Nietzsche." He wrote a multivolume study of American English, but became famous for his iconoclastic reportage for the *Sun*. Mencken's sardonic wit helped heighten public interest in the trial and the issues it raised. His readers were greeted with items like this:

July 10. The trial of the infidel Scopes, beginning here this hot, lovely morning, will greatly resemble, I suspect, the trial of a prohibition agent accused of mayhem in Union Hill, N.J. That is to say, it will be conducted with the most austere regard for the highest principles of jurisprudence. Judge and jury will go to extreme lengths to assure the prisoner the last and least of his rights. He will be protected in his person and feelings by the full military and naval powers of the State of Tennessee.

July 15. The witnesses for the defense, all of them heretics, began to reach town yesterday and are all quartered at what is called the Mansion, an ancient and empty house outside the town limits, now crudely furnished with iron cots, spittoons, playing cards and other camp equipment of scientists. Few, if any, of these witnesses will ever get a chance to outrage the jury with their blasphemies.

Mencken was not exaggerating: Many of the scientists brought as defense witnesses were never allowed to testify. Darrow's team tried to argue that there was no conflict between Genesis and evolution, and attacked the literal interpretation of the Bible, as well as Bryan's limited knowledge of science and of other religions. The defense contended that the Bible should be preserved in the realms of theology and morality but kept out of courses dealing with science. And since the chief witness for the defense was a scientist whom the judge had barred from testifying, the defense concluded by declaring that the prosecution's "duel to the death" with evolution should not be fought in such a one-sided way.

Dudley Malone, a lawyer for the defense, commented: "After all, whether Mr. Bryan knows it or not, he is a mammal, he is an animal and he is a man. . . . There is never a duel with the truth. The truth always wins and we are not afraid of it." On the sixth day of the trial, the judge ruled that the defense's testimony on the Bible was irrelevant. When the defense asked who could be questioned as an expert on the Bible, Bryan said that *he* was such an expert. Thus, with the court's permission, the prosecutor became an expert witness for the defense.

Darrow questioned Bryan on the story of Jonah, on the account of Earth's standing still, on the supposed date of Earth's creation (4004 B.C.), on the story of Adam and Eve, on how Cain got a wife. Bryan answered that he believed in the miracles described in the Bible, did not know how old the planet was, and did not know how Cain got a wife.

The defense was allowed to enter into the record written testimony by experts who were not allowed to testify in person. These included Dr. Maynard Metcalf, a zoologist from Johns Hopkins University, who argued: "There is no conflict, no least degree of conflict, between the Bible and the fact of evolution, but the literalist interpretation of the words of the Bible is not only puerile; it is insulting, both to God and to human intelligence."

Another expert was Wilbur Nelson, state geologist of Tennessee: "In connection with evolution, it is especially of interest to note that the relative ages of the rocks correspond closely to the degrees of complexity of organization shown by the fossils in those rocks. The simpler organizations being found in the more ancient rocks, each type of organism being more and more complex as we come nearer to the present day, man and his fossil and cultural remains being no exception. It, therefore, appears that it would be impossible to study or teach geology in Tennessee or elsewhere without using the theory of evolution."

Dr. Fay-Cooper Cole, an anthropologist from the University of Chicago, wrote: "Only a few points relating to man and his history have been reviewed, but enough has been said to indicate that the testimony of man's body, of his embryological life, of his fossil remains, strongly points to the fact that he is closely related to the other members of the animal world, and that his development to his present form has taken place through immense periods of time. From the above it seems conclusive that it is impossible to teach anthropology or the prehistory of man without teaching evolution."

For weeks, the trial dominated the front pages of newspapers in the United States and received coverage around the world. Despite the efforts of the team led by Darrow, Scopes was convicted and fined $100. Bryan offered to pay the fine. Throughout his trial, John Scopes never took the stand in his own defense—it was unnecessary, since he had never denied teaching evolution. At the end of the trial, though, he made one statement. "Your Honor," he said, "I feel that I have been convicted of violating an unjust statute. I will continue in the future, as I have in the past, to oppose this law in any way I can. Any other action would be in violation of my ideal of academic freedom—that is, to teach the truth as guaranteed in our constitution of personal and religious freedom. I think the fine is unjust."

Mencken, in his last dispatch on the Monkey Trial, observed:

> July 18. Darrow has lost this case. It was lost long before he came to Dayton. But it seems to me that he has nevertheless performed a great public service by fighting it to a finish and in a perfectly serious way. Let no one mistake it for comedy, farcical though it may be in all its details. It serves notice on the country that Neanderthal man is organizing in these forlorn backwaters of the land, led by a fanatic, rid of sense and devoid of conscience. Tennessee, challenging him too timorously and too late, now sees its courts converted into camp meetings and its Bill of Rights made a mock of by its sworn officers of the law. There are other States that had better look to their arsenals before the Hun is at their gates.

On appeal to the Tennessee Supreme Court, the conviction was overturned on technical grounds, but the court upheld the state law forbidding the teaching of evolution.

AN OCEAN AWAY FROM TENNESSEE, Teilhard had a particular interest in the trial and the questions it considered. For he was battling against similar powerful forces bent on denying evolution and imposing a strict, literal interpretation of scripture. But his situation was graver. He was, and was determined to remain, a member of a Catholic religious order that strictly adhered to its interpretations of the Bible and had zero tolerance for dissent. Perhaps he would have gladly paid $100 for the right to be left alone. But this was not his destiny. His punishment for advocating evolution was exile, and as the years went by, this state of exile would become more permanent.

THE WORLD'S REACTION TO PEKING MAN—once he was discovered—would be much more enthusiastic than its reaction to the discovery of *Australopithecus*, which came just before the Scopes trial. The trial in Tennessee had increased worldwide interest in evolution, and people were now looking to science to produce incontrovertible and definitive proof. They were looking for the body that the Scopes trial had failed to deliver—the true missing link between apes and humans: a well-preserved skeleton of a creature that had attributes of both humans and less advanced primates and that was able to make tools and fire.

Chapter 9

THE EXILE

As the Scopes trial progressed in the United States, Pierre Teilhard de Chardin was becoming a major embarrassment for the Society of Jesus in Europe, and had to be dealt with. The Vatican was fully aware of his position: a Jesuit priest who was also a prominent scientist commanding a wide audience among Europe's intellectual elite, and who did not espouse the Church's views on concepts it held sacred and unchallengeable—original sin, Adam and Eve, the Garden of Eden, the Fall, and creation. A challenge of these dogmas could not be tolerated. The Vatican, therefore, pressured the Jesuit superior general to do something about Teilhard, and Ledochowski did not need more than a nudge. The Jesuit leaders in Rome also felt that Teilhard could not be trusted. He had crossed them before, with his papers and lectures exploring the theory of evolution, and there was no guarantee that he would not do it again. As an ordained priest, he could not easily be defrocked or excommunicated for what he had done. Teilhard was entitled to stay a Jesuit priest—as he clearly desired—but a way had to be found to stem the flow of his lectures, conversations, and writings, whose contents the Church considered dangerous.

Even with his mounting problems with the Jesuits, Teilhard remained in Paris throughout the rest of 1925 and into the following year. This was a prolific period for him. He produced excellent scientific results with

Boule and his other colleagues, using his findings from the Ordos, and he developed further his philosophy of the universe, based on evolution. He saw absolutely no conflict between this theory and a Christian belief in God. To Teilhard, the final outcome of evolution was his Omega Point: the point of convergence of all lines of evolution.

Despite the fact that Teilhard had signed the six points at the insistence of his superiors in Rome, the Jesuit fathers still wanted him out of France. They did not believe that he would refrain from making public statements about evolution. Since Teilhard was a leading scholar, people listened to him; and since he was also a Jesuit priest, what he said reflected badly on the Church. The authorities in Rome kept up the pressure on him, and in the summer of 1925, just as the Scopes trial was heating up in the United States, Teilhard was embroiled in a bitter conflict with his superiors about the same issues being raised in the trial in Tennessee. Even the language of his confrontation with the Church echoed that of the Monkey Trial in the United States: Adam and Eve, original sin, creation, and evolution.

To Teilhard, as to Clarence Darrow and the Scopes defense team, there was no conflict between the Bible and evolution. To William Jennings Bryan and the prosecution, as to the Vatican and the Jesuit authorities, the two worldviews were as far apart as possible, and they were irreconcilable. At about the time the evolutionists lost their case in Tennessee, Teilhard gave up his fight to remain in his home country. He told Auguste Valensin that all he wanted now was just six months to put his affairs in order, and then, reluctantly, he would be ready to return to China and rejoin Licent.

Teilhard was anguished and frustrated; above all, he felt that he was not being treated fairly by an organization whose laws he had sworn to obey—which, in his own mind and heart, he did obey. But equally, he felt he could not betray his conscience or what he viewed as his role in the world: that of a devoutly religious man and at the same time a devoted scientist. "What vexes me," he wrote, "is that I believe distinctly that what is demanded of me is a *useless sacrifice*: They imagine that geology corrupts me, while for me it is calming and distracting.—They also think that my teachings are dangerous, while they are purely technical.—They have little confidence that I would come smoothly to the outward attitude that they demand of me—or at least, which is likely, they dread my 'general influence' within a post that is too much in the public view."

As Teilhard later explained to a friend, Ida Treat, geology was his "root," ever leading him forward by its vigor and liveliness to "questions of humanity: unification, exploration, and organization (above all, psychological) of the human level." He resolutely felt that he could never live outside this science.

Teilhard understood well his situation vis-à-vis the Church. From letters and reported conversations we know that he was aware that the order held that he had made an irreversible error, one that had caused irreparable damage to its reputation. The problem was one of lost trust between Teilhard and the Catholic Church.

It is important to understand that there were various levels to Teilhard's conflict with his order and the Church. The Jesuits were, and are, an order with a degree of independence from the Vatican. They are governed separately from the Vatican, with their own head. The order is concerned with how it is viewed by the pope, the cardinals, the Holy Office, and the Vatican as a whole. Teilhard's teachings as a leading scientist and Jesuit could therefore cause great embarrassment to the order and precipitate problems between the Jesuits and the Vatican. French Jesuits have always tried to maintain a degree of independence from the Jesuit establishment in Rome. Thus Teilhard might have expected support from his provincial in Lyon. But his views were so outrageous for a Jesuit, or for any Catholic priest, that he did not enjoy the support he needed either at the level of his order in Rome or at the provincial level in France; his provincial in Lyon offered him little protection.

In the secular world, however, Teilhard was gaining ground, notably in the French intellectual milieu. Through his work with Boule at the museum, he met the poet Paul Valéry, and the two became good friends. Léontine Zanta, a noted feminist woman of letters, entertained him with other intellectuals at her salon. And he would frequently meet with the physicist Maurice de Broglie, whose brother Louis contributed much to quantum theory, and discuss evolution with him. Among the Jesuits and other Catholics, Teilhard had several staunch supporters of his views. They included Auguste Valensin, as well as Édouard Le Roy, who had written favorably about evolution and was a great admirer of Henri Bergson and his work.

When, in late 1925, the Jesuit authorities determined that it was too risky to have the controversial priest remain in Europe, and ordered Teilhard to return to China, he accepted his fate without complaint. He had been readying himself for the inevitable decree. He spent his remaining time in Paris laboring harder than ever in the museum, perhaps feeling this was his last chance to work with Boule on the questions that so engrossed him. And likewise, he continued to occupy himself with problems of religion, lecturing occasionally during the months before his departure.

In April 1926, Teilhard boarded the *Angkor* at Marseille for what he knew was a voyage to exile in China. Thereafter, whenever he returned to Europe, it would be as a visitor, one who was there with the permission of the Church and who would have to depart as soon as he was ordered to do so.

Teilhard's poetic letter to his cousin Marguerite Teillard-Chambon reveals how he felt on leaving France:

Aboard the *Angkor*, April 26, 1926

For the third time since 1906, I pass in front of Bonifacio [Corsica]. Today the sea is gray and rough, not the indigo lake I traversed while journeying toward Egypt, there to find my first awakening. But now my vessel is stronger, and will travel farther, than the one that carried me then to the same Orient. I have certainly grown older, since three years ago, since eighteen months ago. Ideas are no longer born in me with the same exuberance, the same ecstasy, they once had. Those ideas, that fecundity, doesn't last for long in one human life. But on the other hand, deep down, I do not believe I have changed. More coldly, almost without joy, it's the same possession of the world that I seek. Twenty-five years of experience have taught me how to better define nature and the sense of charm I once saw vaguely floating in Matter. Now I see more clearly, I adhere more firmly. But I feel less. It's the same charm, without its charm, that I now pursue. Light once twinkled for me from every surface I saw, and I took immediate pleasure in everything. Now it seems extinguished. The passing film of colors and places now bores me to tears. That which I love I see no more.

The town of Bonifacio has the same name as a number of popes. This point was surely not missed by Teilhard, and perhaps it helped precipitate his mood. His reference to "Matter" recalls the title of his 1919 essay "The Spiritual Power of Matter."

As the ship sailed along the Egyptian coast on approach to the Suez Canal, Teilhard reminisced with melancholy about his arrival at Ismailiya twenty years earlier, when he was twenty-five, and about his enthusiasm and excitement as life and adventure were unfolding in front of his eyes. "Now I think that I could do beautiful things," he wrote Marguerite, "if only I were ten years younger."

Soon after writing his cousin, Teilhard's mood improved. Here was a man who loved the world and people much more than he had realized in his depressed state when leaving France. He had, in fact, much stamina, energy, and lust for life—more than he had gauged he might still have at forty-five. Important years lay before him, not behind. He was too optimistic a person to stay in a depressed state of mind for long, especially since he was about to encounter some of the most fascinating people he would ever meet.

Among the passengers on the *Angkor* was a very interesting couple. It did not take long for Teilhard to discover Henry de Monfreid and his wife, Armgart. They were en route to Obock, their "fiefdom" in Africa, via Djibouti—the *Angkor's* next port of call.

As the ship passed through the Suez Canal and entered the Red Sea one warm, sunny morning, Teilhard was sitting in his usual deck chair writing in his notebook, when he suddenly looked up—he had felt that someone was watching him. In front of him he saw a tall, muscular man with moustache and tanned face, relaxed and confident-looking. They struck up a conversation—and continued to talk every day throughout the voyage. Teilhard and the Monfreids became friends. In fact, with them he forged one of the most enduring friendships of his life. What was so unusual about this friendship was that Teilhard and the Monfreids seemed to have nothing in common. He was a devout priest obediently going to his place of exile, while Monfreid was variously described as a pirate, an arms smuggler, and a drug dealer. His most common moniker was "The Pirate of the Red Sea." He was a mysterious character, a Frenchman who went to Africa because he refused to obey the rules of European society and felt it was his right to trade in any commodity he pleased. He was a free spirit living on the edge, ever involved in shady deals. But his personality was so engaging, and that of Armgart so serene, that Teilhard was seduced by their thirst for life and adventure.

Teilhard was so attracted to this couple that, still aboard the *Angkor*, he confessed to Armgart: "I have full faith in Henry, in what he says about himself; but even more truly, I love you, you and him. I wouldn't imagine asking where or what one may say about him. All of that doesn't interest me at all." Theirs was a fortunate meeting, for seeing how this couple lived, without a care and bound by no rules, freed

Teilhard's own spirit and liberated his tormented soul. Henry, for his part, swore that even death would not end their friendship.

The *Angkor* docked at Djibouti in May, and it was time for Teilhard to say good-bye to his new friends. The Monfreids invited him to visit them on their estate as soon as possible; he would accept the invitation, perhaps even earlier than he expected to.

His spirits lifted, Teilhard could now enjoy the rest of his voyage, stopping at Aden, Colombo, Singapore, Saigon, and finally Shanghai. Since the ship stayed at Saigon for a few days, Teilhard took a train through the jungles to Hanoi. A wide-ranging fire was burning these ancient forests, destroying everything in its path. It was a horrible spectacle; Teilhard saw flaming trees, and terrified animals fleeing the jungle. He took fire to symbolize all the forces that nature can unleash on unsuspecting living creatures, and was awestruck and saddened. "For some years now," he wrote, "I have been distressed to see the countryside retreat and magnificent animals (buffalo, elephants, tigers) disappear."

Once the ship reached Shanghai, Teilhard went to Tientsin on a small Japanese ship, since the trains were not running. On June 10, he entered Licent's laboratory, where his confrère put him to work cleaning and identifying fossils he had recovered from Mongolia in 1925, at the same site he and Teilhard had explored three years earlier.

Soon Licent decided it was time for another expedition. He hoped to travel with Teilhard west along the Yellow River to Gansu Province and into the Tibetan Plateau, at a height above 6,500 feet, to look for new fossils. The pair began the elaborate preparations for this new expedition, arranging for passage permits and supplies, and assembling a caravan. They took a train from Peking south to the Yellow River, and from there joined the east-west railway line to travel far into the west. "Licent had thought that from Lantcheou [Lanzhou], in the Kansou [Gansu], we could continue to Chinese Tibet," Teilhard wrote

Marguerite. "But there we would be stuck without any communication for several weeks."

Teilhard kept writing on the journey. "From Shanchow, we had to continue by foot for more than a week before finding the mules we needed. The haggles over them have been incessant. Now we are finally well on our way to Lantcheou (20–25 days). Until Shanchow, the route is well-passable without much hindrance."

China was by now a far more dangerous place than it had been when Teilhard had previously been there. The Chinese general Chang Tso-lin, who had ruled Manchuria since his appointment as inspector general in 1918, had recently descended with his troops and was warring against two other generals, Feng Yu-hsiang and Wu Pei-fu, for control of northern China. The Chinese Nationalists had embarked on a major offensive against these warlords in the region the two Jesuits had hoped to enter. Licent then chose a route that would take them on the old Silk Road going west. But about five hundred miles into their journey, the caravan ran into trouble. They encountered one of General Feng's regiments, which refused to let them proceed west.

The Jesuits found the road blocked in front of them, three miles before the town of Sinanfu (Tsinanfu). As Teilhard described it in a letter to Marcellin Boule in Paris, he and Licent were caught "in the line of fire. We noted that it would be impossible to go around this front." They discovered to their chagrin that "the valley of Wei-ho is actually a veritable trap. It is impossible to reach Gansu this year." Licent, shrewd in his calculations and experienced in the ways of the Chinese, simply changed course again. They now turned north, toward the highlands of Shanxi Province. There they stayed for some weeks digging for fossils.

As soon as they returned to Tientsin, Licent decided it was time to revisit the Mongolian desert, the site of their richest paleontological finds. They came back from their new Mongolian expedition laden with seven large cases of fossils for the museum, and Teilhard set to work again, writing papers about the geology and paleontology of China. In a letter to the Abbé Gaudefroy, he told of the finds the pair had recovered from this most recent dangerous and difficult journey, during which they once again had to evade highwaymen and make do with very few comforts and little food. What they had discovered this time, he wrote, was "a vast lakeside formation, containing the earliest recognized Pleistocene fauna in China. We found fauna of great variety—but unfortunately nothing recognizably human." An exile he may have been, but Teilhard could still enjoy as many adventures as he wanted, and no one hampered his work as long as he was far from Europe.

Once he had made good progress on cleaning, classifying, and analyzing the fossils he and Licent had collected on their expedition, Teilhard made a special effort to visit Peking often. If he could not be in Paris, at least Peking offered something closer to the cosmopolitan intellectual environment he craved. On his previous stay in China, he had been delighted to find a vibrant community of both Chinese and foreign scholars in the city, and since he was now forced to live in the country indefinitely, he wanted to make the best of the situation and take advantage of the social, professional, and scientific connections Peking could provide. Teilhard renewed his friendships there, and made new ones with Americans, British, French, and others who made up a thriving social group.

Teilhard sought out Davidson Black at Peking Union Medical College, and Johan Andersson, whom he knew through their joint ties with the China Geological Survey, and other researchers he had met. With these renewed acquaintances and after various scientific meetings in Peking, Teilhard was invited by Andersson to participate in the gala event on October 22, 1926, where the Swedish royals were welcomed to China. Teilhard was excited about Andersson's renewed excavations at Zhoukoudian. Like the Swedish geologist, he felt that great treasures lay hidden in the caves of Dragon Bone Hill, and he was elated to hear from his colleague after the banquet that the prince had made the decision to support the temporarily stalled project.

Teilhard returned to Peking a number of times afterward and attended several professional events that attracted scientists from around the world. He shared ideas and research results with these experts, all of whom were aware of his reputation.

Shattering his contentment was the news he received from France: The Jesuit authorities in Rome had revoked his appointment at the Institut Catholique. Teilhard still maintained his association with the Museum of Natural History in Paris, but without a position for him at the Institut, his situation in Europe was precarious. In essence, he had been blocked from returning. When he received this news, he wrote to a friend that his roots were in Paris and would always remain there— for how could anyone dig them up? Deep down, he confessed, he did not understand the East. But great scientific opportunities lay here, and whatever the Jesuits could do to his standing in Europe, he was now here and ready to participate in a major professional adventure.

Out of acute frustration, sadness, and despair about his predicament—and in a sincere attempt to appease his church—Teilhard began writing a book, which he titled *Le Milieu Divin* (*The Divine Milieu*). He often described it as "a little book of piety," and it was centered on his global view of religion and humanity's place in the universe.

While working on the book, Teilhard wrote constantly to his superiors in Rome and his provincial in Lyon, requesting a return to France. He proposed that he become a traveling priest, regularly spending eighteen months working on paleontology in China, followed by six months in Paris analyzing his findings. He reminded his superiors that he had scientific obligations to Boule at the museum, and that he needed to dispense with these obligations. While he waited for a response from the Jesuits, he finished his *Milieu Divin*. He was proud of the book, and he hoped to receive permission from the order to publish it.

After months of considering his request for leave, the Jesuits relented and granted Teilhard permission to travel to France. His friends Valensin and Le Roy had intervened on his behalf with his provincial, who in turn made requests of the authorities in Rome. Teilhard was happy to go home, and on August 27, 1927, he boarded a ship in Shanghai.

Once in Paris, he pushed hard for permission to publish *Le Milieu Divin*, but got nowhere with the Jesuit censors in Rome. With its geocentric approach, in which Teilhard attempted to reconcile the love of God with the love of the planet, the book was viewed as irreligious—perhaps even pagan. This was ironic, since he had written it in part as an effort to please the Church. *Le Milieu Divin* did not include any challenge to original sin, and it dealt with traditional Christian symbols: the cross, baptism, and the eucharist. But its devotion to nature, to the planet, and to all creation—as well as the inherent notion of growth and development, which hinted at evolution—made the Church suspicious. In fact, authorities were so worried about the text that an unrevised copy of it had, under mysterious circumstances, "disappeared" from Teilhard's possession and materialized in Rome, where it was examined.

In the meantime, some of Teilhard's friends, in particular Valensin and Le Roy, were incorporating into their lectures elements of his terminology and ideas, and attributing them to him. This further infuriated his order, and in 1928, a Jesuit traveled to Paris to warn Teilhard again that unless he changed his ways, he would be banished to an even more remote location than China, and would not be allowed to continue his scientific work.

With these conditions, Teilhard realized, he was better off in China, where he was not under constant scrutiny. He would return to his place of exile, where he had a good support group and an immensely important project in paleoanthropology.

Teilhard spent his remaining months in France socializing, renewing friendships with those Jesuits who loved him, visiting his extended family, and traveling. He lectured to students from the École Normale

Supérieure in Paris, using the ideas of *Le Milieu Divin*, and gave them copies of his manuscript. He wrote a paper, "The Foundation of the Idea of Evolution," and, at an intellectual retreat in France, gave a copy to the Abbé Bruno de Solages. The abbé was impressed with the way Teilhard was able to reconcile Christian belief with evolution, and he interviewed him at length about this subject. An intelligent and progressively minded cleric, Solages understood Teilhard's ideas and saw that there was nothing subversive or anti-Christian in his embrace of evolution—that Teilhard may have found a way to reconcile science with religion. Solages became one of Teilhard's most ardent supporters, and a powerful defender of his views.

Bruno de Solages was a courageous, steadfast man. During World War II, he would serve in the Resistance and be imprisoned in a Nazi concentration camp. As a monsignor, he would serve as rector of the Institut Catholique in Toulouse and as prelate to the pope. He would be Teilhard's staunchest defender in Rome. The Catholic Church could not have been more irritated by Teilhard's winning agreement for his theories among its prominent members. Teilhard left Paris for his family estate in the Auvergne, where he stayed with his favorite cousin, Marguerite Teillard-Chambon, then continued southwest. He visited several caves in the Pyrenees in which prehistoric art had been found. He was accompanied by his friend the Abbé Breuil, whom the French still call "The Pope of Prehistory" because of his many discoveries and authentications of Cro-Magnon cave art, and by a friend with whom Teilhard had served in the war, Max Bégouën. They stayed at the château of Bégouën's father, which was located near several prehistoric caves in the Ariège region of the Pyrenees. In this bucolic setting, Teilhard started writing the first few pages of a new essay, which he called "Le Phénomène Humain" (The Phenomenon of Man). Later, he would write a book with this title.

On November 7, 1928, Teilhard left Marseille aboard the *Chantilly*, destined again for China. On the way, he stopped in Africa to visit the

Monfreids at Obock. He was traveling with a geologist, Pierre Lamare, and together they disembarked at Djibouti.

HENRY DE MONFREID had left France in 1911, in his early thirties, to seek his fortune on the Red Sea. He established himself at Obock, north of the city of Djibouti, in then French Somaliland, and made a great deal of money in the pearl trade. He learned Arabic, and found himself at home in this rough and lawless part of Africa. Once he understood that Somalian and Ethiopian warlords had an insatiable appetite for weaponry, he became an arms trader as well, and further increased his fortune and his influence around the Horn of Africa. From there, it was not a big switch to the opium and hashish trade, which proved more profitable than the weapons. But Monfreid had an interest in science, too, and he had chanced upon some fossil remains at a location inland from his coastal home, in the Great Rift Valley—the same area where the most important discoveries of African fossil hominids would be made throughout the twentieth century by the Leakeys, Donald Johanson, Alan Walker, and others.

Teilhard and Lamare were met warmly by the Monfreids at their sprawling seaside estate. Lamare had met Monfreid when he had visited Africa some years earlier—Monfreid was a famous Westerner here, and European travelers to the region often sought him out. Teilhard and Lamare were entertained by the Monfreids, and by many local people who came to visit, curious to see the scientist-priest and the geologist. Monfreid and his guests shared a penchant for exploration, and they made daily forays into the interior.

One day Monfreid took his two visitors on a trek to a mountain to which no other white person had gone. The natives knew, feared, and revered "the pirate," and their respect intensified when they saw him accompanied by two other tall white men, all three in elegant safari clothing. During one of their hikes, Teilhard was stung by several especially venomous African wasps whose nest the three had disturbed. Monfreid, who was stung by one wasp and was in great pain, had tremendous respect for Teilhard, who bore the pain of his many stings without a word.

Teilhard wrote Marcellin Boule:

Obock, November 26, 1928

Dear Maître and friend,

I write you from Obock (60 Km from Djibouti, by sea), where I have now been twelve days—partly because my friend Monfreid has work to do, and partly because we are stuck here because of the caprices of the governor of Djibouti (Chapon-Boissac, or rather, it seems, his wife). Due to jealousy or rancor, this man believes that my tour among the Dankalis [a local tribe] would allow de Monfreid political agitation (!) or traffic in arms (!!); and for this ridiculous reason, he has sent an entire expedition here to keep an eye on us. This affair has become the laughingstock of the colony, and has assured me a personal popularity among the Dankalis. But so that I cannot just as well pass into a formal rebellion, I am being prevented from going to explore a massif, 15 kilometers from here, where the Dankali chiefs are awaiting us with open arms!

Eventually, Monfreid, Teilhard, and Lamare entered Ethiopia, slipping illegally across the border. Monfreid had nothing but contempt for the French colonial authorities, and the natives of the region revered him as a hero; with their help he could cross any border he wanted. For a month, the three men explored the rocky deserts and then headed inland to the Harar Plateau. This was a wild land, like the biblical Garden of Eden, rich in bananas and other fruits, birds of exotic varieties, antelopes, and baboons. At Christmas, Teilhard found

a Capuchin monastery in the jungle in which to say Mass. They continued farther south to the Errer Valley, where Teilhard explored Paleolithic sites.

On December 28, 1928, Teilhard wrote to Marguerite from Harar
about visiting the Afar region. His intuition about the origin of human
ancestors was astonishingly accurate: While he could not have known
it at the time, it was in this region that, some forty-five years later,
Donald Johanson would discover the most famous hominid fossil in
history: the *Australopithecus afarensis* known as "Lucy."

Traveling in this part of Africa, far from his enemies in Europe, the
beleaguered priest found his freedom; he felt liberated and young
again. In February 1929, Teilhard was on board ship again, this time the
André Lebon, headed for Hong Kong, from where he would take
another ship to Shanghai, to continue to Tientsin. Before leaving
Africa, he sent five cases of animal fossils to Boule's laboratory at the
museum in Paris. According to his biographer Jacques Arnould, Teilhard once brought Monfreid opium from China—"for his personal
use." He also saved Monfreid by intervening on his behalf with authorities who trapped him while he was trying to cross into Chinese
Turkestan, where he was supposed to pick up a shipment of hashish.
And when the Monfreids' son Marcel was alone in Saigon and faced
business difficulties, Teilhard visited him and offered his help.

On board the *André Lebon*, Teilhard wrote notes in which he criticized the Church for its inability to reconcile itself with the modern
world and embrace new ideas and science. It was locked in its ancient
position of "verbal theologizing," and could not see that the world was
moving forward. "The time has come," he wrote, "for us to save Christ
from the clerics, in order to save the World." Teilhard had become a
true rebel.

Chapter 10

THE DISCOVERY OF PEKING MAN

Upon his return to China, Teilhard immersed himself in the paleo-anthropological project. He had first visited Zhoukoudian in 1926, and was enthusiastic about the work being done there. While he initially had doubts that early hominids would be found, he was now quite convinced by the preliminary findings—as were Davidson Black and others—that this place held potential for yielding fossils of the missing link.

In Teilhard's absence, Johan Andersson's work had taken a promising turn: Peking Union Medical College had agreed to support his excavations. This happened when the Rockefeller Foundation, which funded the college, reversed its previous course after learning that Prince Gustaf had announced that he would generously finance hominid research in China. The Foundation no longer opposed Davidson Black's preoccupation with this endeavor, and provided its own funding for the Zhoukoudian project.

To implement its renewed involvement with the fossil search, the Foundation established the Institute of Human Biology and Cenozoic Research Laboratory, to be housed within Peking Union Medical College. The Institute and its laboratory would be where fossils that were discovered would be kept and analyzed. Black was appointed to head the Institute.

On March 27, 1927, exploration of the Zhoukoudian caves resumed in earnest and with full funding. The members of the group assigned to do the fieldwork were housed at a camel caravanserai called the Liu Shen Inn, the only accommodation in this rural part of China. The team lodging here included Birger Bohlin, the Swedish paleontologist hired for this project, and a number of Chinese geologists. Sixty laborers were hired from among the locals. Black, who supervised the project, remained in Peking, but he often traveled to Zhoukoudian.

With this support of the operations at Zhoukoudian, the Rockefeller Foundation was competing with another American institution keenly interested in the search for hominid fossils in China: New York's American Museum of Natural History. The museum had planned to send its own team of scientists, but Andersson, with a healthy research budget now coming from two sources, managed to consolidate his hold on Zhoukoudian. When the museum's American team arrived in China, they could obtain permission to excavate only in Mongolia, where they found dinosaur bones but no human remains.

By the end of the 1927 digging season, the team at Zhoukoudian had uncovered another humanlike tooth. Davidson Black was ecstatic. He thought that the find confirmed his confidence in naming the new species *Sinanthropus pekinensis* (Chinese Man of Peking). He now applied to the Rockefeller Foundation for more money—and got it. His operation was well on its way.

Licent knew that the scientific community in Peking had very high regard for Teilhard, but not for him, and this made him envious of his colleague. When Teilhard reached Tientsin in early March 1929, an irreconcilable difference of opinion between him and Licent became evident. To Teilhard, each ancient bone told a story—about life, about the past, about evolution, about God's creation. To Licent, the fossils he collected were simply things to be labeled and displayed. There was a disagreement in how the two men viewed the purpose of the museum.

Licent also believed that the museum was a French outpost in a foreign land—its collections were not to be shared with the Chinese or with other Westerners. He liked to work either alone or with Teilhard, and he wanted to keep his findings in the museum, not share them with others. It was "his" museum. Licent had been living in China for many years, and had long before formed his isolationist habits and opinions: He did not trust the Chinese, and he had few Western friends or allies.

Teilhard, in contrast, believed that knowledge and science belonged to humanity and were to be shared by all. An example of such international cooperation and the unity of science was Zhoukoudian, where scientists of many nationalities collaborated to uncover human ancestry. Teilhard was social, comfortable with people of all backgrounds and nationalities, and he reveled in conversations and shared experiences. He always looked forward to his trips to Peking, where he would meet people and exchange ideas. The disparities in style and outlook between Teilhard and Licent were so extreme that an eruption of conflict between them was just a matter of time.

The China Geological Survey oversaw the Zhoukoudian project, and the director of the enterprise was now the Belgian-trained Chinese paleontologist Weng Wenhao. Davidson Black was named honorary chairman, and he asked Teilhard to assume the role of honorary advisor on vertebrate paleontology. Teilhard accepted the role—despite vehement objections from Licent, who wanted him to remain in Tientsin—and formally joined the international effort at Zhoukoudian.

From then on, Teilhard would spend even less time in Tientsin with the increasingly unpleasant Licent, and would take every opportunity he had to visit Peking, where he was given the use of a laboratory at Peking Union Medical College not far from Davidson Black's own laboratory. Teilhard's new position with the China Geological Survey was the best appointment he could hope for: It involved him directly in a major project that combined his interests in geology, biology, anatomy, paleontology, and anthropology. His years of education, field preparation, and hard work, and his enthusiasm for searching for fossils made him the perfect scientist and scholar to join this multinational effort.

In late March 1929, Teilhard wrote to Marcellin Boule about the many finds already made at Zhoukoudian:

I have been coming to Peking to renew contact with numerous friends. One of my first visits was for the *"Sinanthropus,"* about which I had learned, with astonishment, already in Shanghai, that they had found quite a number of pieces at Chou-kou-tien (in ultra-satisfactory stratigraphic conditions). Black now possesses a score of isolated teeth; a large fragment of the mandible of an adult; a complete mandible of a juvenile (that is, the right branch complete, and the left branch until the first premolar); the parietal of an adult; several fragments of a skull (small) of a juvenile. . . . Now, I want to tell you that these pieces have made me greatly "excited" [in English in original]. Black is going to publish soon (all the drawings and photos are ready), and he will start with an article in *Nature.*

The next month, Teilhard made his first visit to Zhoukoudian since his return to China. He was surprised to see how much more extensive the excavations were since his previous visit. What used to be a single trench some thirty square feet in area and seventeen feet deep had grown into a sprawling digging site 400 feet long and 120 feet across at its widest point, in which explosives were used to pulverize large rocks, and many workers were removing huge piles of earth and debris.

According to Jia Lanpo, the youngest of the Chinese excavators at Zhoukoudian in 1929, Birger Bohlin was absent from the excavation site that year. Jia wrote:

After conferring with Davidson Black, the head of the Geological Survey of China, Weng Wenhao, put Pei Wenzhong in charge of the Zhoukoudian project. On a glorious April day that year, Pei, Black, Teilhard, and Yang [Zhongjian] arrived at the site for a conference on further excavation plans. It was decided then that the target was to be the middle section of the areas which had been dug in 1927 and 1928, from the fifth layer on down to the very bottom of the fossiliferous deposits.

In previous years, Bohlin, the geologist Li Jie, and Yang Zhongjian shared the administrative duties at the site. But this year, Pei was assigned all responsibility. He later confided to Jia that he was seized by melancholy after Black, Teilhard, and Yang left the site, because he was now alone as head of the operation and was faced with the unpleasant task of excavating the extraordinarily hard fifth layer, which had proven resistant to explosives.

This very dense fifth layer, however, was not especially thick, and within a few months, Pei and his workers were able to break through it to lower ground. The sixth layer yielded animal fossils, which were encouraging, and once the workers reached a seventh layer, fossils appeared with even greater frequency. As Pei described it, one day they found 145 jawbones of thick-jawed deer—fossils that were both plentiful and in good condition. The excavators even found complete pig and buffalo skulls, and deer antlers. They continued to an eighth layer, where they found more animal fossils.

Then, in the ninth layer, the excavators found several teeth that looked human and fragments that resembled portions of a crushed skull. But the enterprise still awaited its great discovery. Early man had been here, Pei and other trained researchers knew, and had left traces of his dwelling at Dragon Bone Hill, but where were *his* remains? Where was the definitive proof that everyone felt had to be hidden somewhere in these caves? The scientists needed the most convincing proof an anthropologist must have: a well-preserved skull.

The particular location at which the teeth and encouraging fossil fragments were found was designated "Ape-Man Locus C," reflecting the

A recent (2005) view of the cave at Zhoukoudian where the first Peking Man skull was found. Photograph by the author.

scientists' hopes. The digging stopped for some time as teams changed. It started again on September 26. As the fossil deposits became scarcer, the excavated area narrowed. Soon there was very little space in which to work at this location—no more than a few men could stand there at any one time. Pei inspected the space and was about to declare that they had reached the bottom, when suddenly he saw a crack in the southern side of the site. Using a rope, he estimated that it led down to a depth of forty meters (about 130 feet) below the surface.

Once it was opened up by the excavators, the space he had just sighted would be named "Ape-Man Cave." It was a narrow and deep hollow, and the researchers named it for what they now hoped would be uncovered here: the skull of a human ancestor. Pei and a worker were lowered by ropes through the fissure and found themselves inside the cave. They stood there in amazement, ropes tightly fastened to their waists. It was by now November, and fieldwork was supposed to stop for the winter. But Pei and his crew found many fossils within this newly opened cavity, and decided to continue to explore it.

Pei, who had been trained for his job on site, was inexperienced, but highly motivated and enthusiastic. Davidson Black once referred to him as "a corking field man." Just when he was supposed to shut down the operation and go home for the winter, as the nearly frozen earth was extremely hard to excavate, Pei resolved to stay. He now had a new and promising cave to explore, and he would push forward at any cost. He gave the order to keep working at an increased pace.

And then, after so many months of hard and often frustrating work, they found their prize, at a little after four p.m. on Monday, December 2, 1929. Wang Cunyi, who later became a technical advisor to the Chinese Institute of Vertebrate Paleontology and Paleoanthropology, described the event to Jia Lanpo:

> It was near sunset and the winter wind brought freezing temperatures
> to the site. Everyone felt the cold, but all were working hard at finding

more fossils. There were four people down in the pit, but I can recall the names of only three—Qiao Derui, Song Guorui, and Liu Yishan, now all deceased. I was working elsewhere at the site when the Peking Man skull was found. However, I had seen the place, as all of us had. The large number of fossils attracted every one of us and we all went down to take a look, so I know what it was like down there in the crack. We generally used gas light, for it was brighter. But the pit was so small that anyone working there had to hold a candle in one hand and work with the other.

Jia, a newcomer to the operation, had arrived just in time for the greatest moment of the project: the discovery of a skull. Perhaps because of the cold weather, or because of the late hour of the day, or both, he wrote, there was such a stillness in the air that all one could hear was the occasional rhythmic sound of a hammer, indicating that someone was working.

As soon as he heard that something of interest had been found inside the cave, Pei Wenzhong rushed down through the crack and started working with the four men inside this very narrow cavity to expose the object. He realized what it had to be, and he cried out: "What's that? A human skull!" In the tranquillity of dusk, everyone heard him and hurried over to see what was happening; the excitement at finding what seemed to be the long-sought skull overwhelmed them all.

There was a big discussion about what to do next. Some suggested that they remove the skull immediately, while others argued that since it had been there for so many thousands of years, it could stay another night without harm—they did not want to do anything that might damage this precious fossil. Pei thought that a long night of suspense would be too much for anyone to bear. He decided to remove the skull then, rather than wait for morning. Half of it was embedded in hard clay, and so he worked to free it from the matrix.

His biggest worry was how to inform Black and Weng Wenhao in

Peking about the great discovery. That night he sat down and wrote a letter to Weng, and early the next morning dispatched one of his men to Peking with it. Soon after the messenger left, Pei had a second thought: By the time the man arrived in Peking, it would be late in the evening; he wanted the news to get there faster. Pei went to the telegraph office in the village and sent Black a telegram: "Found skullcap—perfect—look[s] like man's."

The next morning Pei wrapped his find in a burlap bag. He had decided to take it to Davidson Black in Peking. Unfortunately, the temperature had risen and a violent rainstorm swept the area, quickly melting the snow and flooding the river that separated Dragon Bone Hill from the road to Peking.

Stuck in this place, Pei and his colleagues continued to work on removing the clay from around the skullcap of Peking Man. It emerged as a more complete skullcap than Dubois had found in Java, with well-preserved brow ridges and an underside. Even after much of the detritus was removed, the skull was still covered in wet clay, and Pei and his helpers spent the night and the following day drying it by the fireside.

In Peking, both the letter and the telegram were hardly believed. On December 5, Black wrote to Andersson, who was then in Sweden: "I had a telegram from Pei from Chou Kou Tien yesterday, saying he would be in Peking tomorrow bringing with him what he thinks is a complete *Sinanthropus* skull!! I hope it turns out to be true." Teilhard was apprised of the discovery, and shared in the euphoria among the group in Peking.

Once the skull was reasonably free of clay, Pei wrapped it in gauze soaked in glue, covered it in plaster, and wrapped the package in a thick blanket and tied rope around it. This was done to protect the find, but also to make the package look like the typical bundle carried by Chinese peasants, and thus avoid an unwelcome inspection. There were several possible checkpoints between Zhoukoudian and Peking, and Pei did not want the authorities to discover what he was carrying, lest they confiscate it.

On the morning of December 6, he forded the turbulent river on foot, holding aloft the bulky package containing the skull of Peking Man. In Zhoukoudian village he boarded a bus for Peking. He would have to pass through a police checkpoint, and this made him nervous. As he later recounted, he had a few other fossils with him and planned to show them to the police and say that the large package held just "more of the same." If he had to open the package, he would try to open only the outermost layer and not expose the skull itself. "If [a policeman] insisted on opening the bundle, the plaster and gauze would have kept [the skull] intact. If he still insisted on taking a look at what was inside, I would ask him to arrest me first."

Fortunately, Pei negotiated the police checkpoint without trouble; his bundle was barely noticed. By dusk he was standing in front of Davidson Black in the Cenozoic Research Laboratory at Peking Union Medical College.

UNTIL NOW, Black's career had been seemingly undirected. He had studied medicine in his native Canada, and then turned to paleontology, which he studied in England. In 1919, he moved to the anatomy department at Peking Union Medical College, hoping to get involved in early-human explorations. After he met Johan Gunnar Andersson, he became so enthusiastic about the Zhoukoudian project that he traveled throughout Europe in 1922, working hard to raise funds for the excavation, with a replica of the tooth found at the site dangling from his pocketwatch chain.

Pei now entered Black's laboratory with the skull of Peking Man in his hands. Black's great moment had arrived.

Black locked the door and spent the entire night separating the fossilized skull from whatever clay remained, on the top part of the skull, using fine dental tools. What emerged after hours of work was the nearly perfectly preserved ancient skull of a hominid resembling

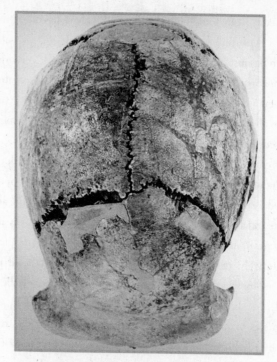

The first photograph of the Peking Man skull, 1929.
Fondation Teilhard de Chardin, Paris.

humans and yet different in some ways. At dawn, Black stared at this object with amazement. A few hours later, he showed the skull to Teilhard and other scientists who congregated in his lab.

Teilhard described this first skull discovered at Zhoukoudian: "as typical a link between man and the apes as one could wish for." Shortly after the momentous discovery, he wrote Boule:

December 11, 1929

Dear Maître and Friend,

Thank you for your letter of 8 November, which arrived here three days ago. . . . In fact, as a New Year's "gift," I would like to announce to you (something that [Jean] Piveteau [a colleague of and coauthor with Teilhard] might have already told you) that at Chou-kou-tien has just been discovered the largest part of a "cerebral" skull of a *Sinanthropus*; *non-deformed*, and almost completely separated from the travertine [massive, layered calcium carbonate deposit] (to which only the edges of the specimen now seem attached).——The piece was just discovered at the lowest parts of the exploration area. . . . It is not yet completely extricated, but one already sees the upper part, from the orbits to the occiput. At least part of the nasal bone is still enclosed in a block of travertine. Perhaps there would be the auditory conduits and also something of the foramen magnum (?). The jaws are missing. As it is, the piece is stunning. Size of the *Pithecanthropus*, approximately. (Slightly larger?)——but distinct frontal and parietal humps (somewhat similar to those on the skullcap of the Nean-derthal). . . . In short, the *Sinanthropus* presents itself, at this moment, as making an extremely precise morphological transition between *Pithecanthropus* and Neanderthal.——One may hope for new discoveries. The search is over because of winter. But we haven't reached the bottom yet, and the point (very fossil-full) where the skull was found has barely been touched.

The international research group in Peking was overjoyed with the discovery of December 1929, and within months the world knew about the astounding find. News reports in many countries focused on the significance of Peking Man.

Teilhard had a key role in the analysis of the Peking Man skull that followed. He helped estimate the age of the find by determining the geological era of the layer at which it was found. He identified and ana-lyzed the stone tools found at Zhoukoudian and presumably used by Peking Man, and interpreted evidence from the cave that led to the paramount conclusion that Peking Man used fire. Teilhard also wrote several articles that established the importance of the scientific find-

The international team behind the discovery of Peking Man at Zhoukoudian in 1929. Left to right: Pei Wenzhong, Wang Hengsheng, Wang Gongmu, Yang Zhongjian, Birger Bohlin, Davidson Black, Teilhard de Chardin, George Barbour. Reprinted by permission of the Institute of Vertebrate Paleontology and Paleoanthropology, Chinese Academy of Sciences, Beijing.

ings at Zhoukoudian, which appeared in the standard Catholic scientific review, published in Belgium, *Revue des Questions Scientifiques* (July 1930), in *Etudes* (July 1937), and in publications of the Institute of Geo-Biology in Peking.

During the next excavation season, in 1930, another complete skull was found in the same cave, as well as a number of teeth. The following year yielded stone tools and strong evidence of the use of fire. In 1932, excavators found a well-preserved jawbone of Peking Man. More skulls, fragments, and tools would be discovered over the next four years, all in excellent condition. Partial and complete skulls were identified with Roman numerals, and by 1936 they would make up a collection of I to XII. In total, five almost complete skulls would be recovered from this level of the excavation at Zhoukoudian.

Today's scientific dating techniques, which use the rates of decay of radioactive elements such as uranium, help determine the age of finds. Peking Man is estimated to have lived at Dragon Bone Hill sporadically from about 670,000 years ago to about 410,000 years ago. So he was on average roughly 500,000 years old—younger than Java Man by 200,000 years, but much older than Cro-Magnon and Neanderthal.

After the discoveries at Zhoukoudian, Davidson Black again embarked on a European tour: this time with casts of the finds and a slide presentation that impressed everyone who saw it. He received a standing ovation wherever he spoke: at universities, research institutions, and public forums. Peking Man was now recognized as a missing link between humans and apes. In 1932, Black was elected to the Royal Society in London in recognition of his work. In Holland, however, Eugène Dubois claimed that his *Pithecanthropus* was the sole missing link and that *Sinanthropus* was simply a Neanderthal. Today we recognize that both finds belong to the species *Homo erectus* and indeed represent what may be considered a missing link.

The large scale of the operation at Zhoukoudian expanded further after the discoveries of the skulls. Better methods were needed to search for more fossils. Pei Wenzhong devised a mechanized digging system, with a cable-and-pulley setup to move baskets of excavation debris from the now sprawling excavation complex.

This was Black's greatest success, but it was also Teilhard's, and it was the pinnacle of the career of the young Pei Wenzhong. The three men's lives would remain intertwined by their participation in one of the most important discoveries of the twentieth century.

Chapter 11

TEILHARD MEETS LUCILE SWAN

Spending most of his time in Peking's intellectual expatriate community, Teilhard was always encountering new people. One of them
was an American artist named Lucile Swan.

Swan was born on May 10, 1890, in Sioux City, Iowa. She attended
an Episcopal boarding school, and when she was thirteen moved with
her family to Chicago. There she began to study art and within a few
years became an accomplished sculptor. In 1912, Swan married
Jerome Blum, a fellow artist, and traveled widely with him over the
next twelve years: France, Japan, China, Tahiti. The marriage showed
signs of stress, and in 1924 the couple divorced.

Swan was an attractive, although not a beautiful, woman. She was of
medium height and full-figured, with an expressive face and a ready
smile. She was sociable and took pleasure in the company of the many
interesting people she met on her travels.

In 1926, at age thirty-six, Swan closed her Chicago studio. She
moved to New York City, where she continued to work as a sculptor,
exhibiting at the Anderson Galleries, which specialized in American
art. Soon she caught the travel bug again and decided to return to a
place she had visited in her earlier travels: Peking.

In 1929, with her friend Betty Spencer, an adventurous woman who also loved to travel, Swan arrived in Peking. They stayed in a large house surrounded by a magnificent garden in the inner city, where most expatriates lived, close to the Forbidden City. Most citizens of Western countries who lived in China enjoyed very comfortable houses; they were served by maids, cooks and cooks' helpers, drivers, gardeners, and other servants. Swan and Spencer were active in Westerners' social life in the city, and regularly hosted and attended dinner parties.

When Teilhard returned to Tientsin after Christmas 1929, Father Licent was outraged. He was angry that Teilhard had stayed away for so long, and saw it as a desertion of Teilhard's duties to the museum. He accused him of "going over to the Chinese" and called him a "coolie." Then he added other insults, "to which, 'in the world' one answers with a blow," as Teilhard wrote to Boule. Teilhard took the abuse in silence—this was the first and only time in his life that he had a serious personal conflict with anyone, and he was too much of a gentleman to fight with Licent. Teilhard had an unusually kind nature. His friend the Jesuit Pierre Leroy described him:

> The look in his eyes when they met your eyes revealed the man's soul: his reassuring sympathy restored your confidence in yourself. Just to speak to him made you feel better; you knew that he was listening to you and that he understood you. His own faith was in the invincible power of love: men hurt one another by not loving one another. And this was not naïveté but the goodness of the man, for he was good beyond the common measure.

Teilhard remained calm and did not respond in anger to Licent, but he decided to leave Tientsin definitively and move to Peking. Since the Jesuits had no house there, he boarded with the Lazarists—a Catholic order founded in the seventeenth century by Saint Vincent de Paul—in a residence near the Forbidden City, close to his laboratory at Peking Union Medical College.

Teilhard spent his free time socializing with Western intellectuals (not speaking Mandarin made it hard, if not impossible, for him to communicate with Chinese who did not speak other languages). He lunched at the Hôtel du Nord with members of the diplomatic corps, and saw friends for tea or dinner. He had so many acquaintances that he rarely had his meals with the Lazarists. The hostesses of the expatriate community found the engaging priest a most desirable guest at their dinner parties; Teilhard was welcome everywhere.

At one of these parties, at the home of the German-American scientist Amadeus Grabau and his wife, Lucile Swan found herself seated next to Pierre Teilhard de Chardin. They were lounging on comfortable rattan chairs around a lacquered rosewood table in a sitting room

Teilhard de Chardin and Lucile Swan in the Western Hills, late 1930s.
Fondation Teilhard de Chardin, Paris.

decorated with wood-and-rice-paper lattice screens. Many of the foreigners Swan had met in China were scientists, and she knew that Dr. Grabau was a geologist, so she guessed that Teilhard also was a scientist. She turned to him and asked: "What kind of an -ologist are you?" Teilhard laughed. He proceeded to tell her about his scientific work and his new position at Peking Union Medical College. She was intrigued. Then Teilhard mentioned that he was a Jesuit priest.

They talked late into the night. Teilhard told her that he saw science and religion as going hand in hand, and explained his mystical view of God, science, and evolution. The deeper he delved into science, he told her, the surer he felt that there was a God.

Swan was charmed. They were strongly drawn to each other, and neither of them understood the consequences of their increasing closeness: She was a freethinking no-longer-married woman, and he was a Jesuit priest—whose vows included chastity.

Lucile Swan decided to extend her stay in Peking. She moved to an abandoned temple that had been restructured as a house, surrounded by elegant high trees, in the west side of the Tartar City in Peking. She supported herself by teaching sculpture to ladies of the expatriate community, and in the afternoons she held tea parties. Teilhard became a regular guest.

WITH THE GREAT DISCOVERIES from Zhoukoudian accumulating in the laboratories of Peking Union Medical College—in total, six almost complete skulls, eight skull fragments, six pieces of facial bone, fifteen mandibles, 153 teeth, seven fragments of femur, pieces of upper arm bone, and various other bone fragments—Davidson Black began studying the fossils of Peking Man in detail. *Sinanthropus pekinensis* was a reality—exactly as Black had predicted. The comprehensive analysis he initiated would be continued by others, notably Franz Weidenreich, who would replace him. What Black started was a complete reconstruction of the Peking Man anatomy, based on the collection of bones and skulls belonging to twenty-five adults and fifteen juveniles recovered from the Zhoukoudian cave. From this inferred anatomy, and from stone tools and animal bones from the site, conclusions were drawn about the lifestyle, nutrition, social structure, and evolutionary significance of *Sinanthropus*.

Peking Man individuals were shorter than modern humans—males were about five-one on average, females four-nine. Harry Shapiro, then at the American Museum of Natural History in New York, made a comparison of *Sinanthropus* with modern humans. He pointed out that except for Pygmies and Bushmen, who average about four-seven to four-nine, modern humans are taller, on average, than Peking Man. Earlier hominids, however, were even smaller in stature. There seems to have been a steady increase in height in the progression from earlier to later hominids. This observation has also been confirmed from the estimated heights of the australopithecines, which were shorter than the younger *Homo erectus*.

One complete collarbone of Peking Man survived, and from it scientists determined that he had relatively wide shoulders. He was a muscular hominid, broad-shouldered and short-legged.

Peking Man had no chin. Below the roots of his teeth, his face sloped backward; and his mouth protruded. He had a low, massive, projecting brow. But he had more of a forehead than his predecessors did. His skull was smaller and flatter than that of modern humans, and he had relatively thick ridges over his eyes. He had a low and broad nose, high cheekbones, and a wide face. One puzzling finding was that Peking Man's skull was much thicker than human skulls.

Imagine meeting one of these extinct hominids on the subway (or on the checkout line or at the mall). How would he look to you? According to Donald Johanson, a Neanderthal would look like a hairy, "wild" man with a stocky body, high cheekbones and thick eyebrows, and a low forehead. We might not even look at him twice, especially if he wore a business suit, trimmed his hair, and had a close shave. A *Sinanthropus*, by contrast, would look very strange indeed—he would seem deformed. And his lack of chin would make him stand out.

How did Peking Man live? According to analysis of animal fossils found at Zhoukoudian, Peking Man ate mammals, birds, fish, turtles, toads, and frogs, as well as seeds and fruits and vegetables. Zhoukoudian was a community of hunter-gatherers who foraged through the countryside around their cave, bringing home anything they could collect or hunt. Stone tools found at the site indicate that these hominids used tools to hunt and butcher the animals they caught, and they cooked them over a fire.

Stone tools had been used by earlier hominids in Africa, including *Homo habilis* (Able Man or Handy Man), dated to 2.5 million years ago, and even australopithecines are believed to have chipped pebbles and used them as tools. The Peking Man tools were more advanced than those of earlier hominids, but not as sophisticated as the later stone tools of Neanderthals and humans in Europe.

One of the most interesting things about Peking Man was that he lived at a location with cold winters. Java Man lived in the tropics, where it was always warm; but Peking Man, to survive the freezing winters of northern China, needed fire. And indeed, scientists found evidence at Zhoukoudian that the Peking Man individuals controlled fire and used it to heat their cave and cook food. Among the finds from the cave were cooked gazelle bones and other charred animal remains.

These hominids were in constant danger from large carnivores. Primates are small mammals in general, and a single hominid was no match for a tiger, hyena, elephant, rhinoceros, or bear. Their survival depended on their staying together as a community, hunting together, and defending themselves as a group. The Peking Man individuals formed a cohesive population in which everyone needed everyone else to survive. Some anthropologists have suggested that the Zhoukoudian cave was actually the den of large carnivores that preyed on Peking

Man, and that the hominid fossils found there were the remains of the individuals these animals caught and dragged to their cave to eat.

Behind this theory was the observation that many of the skulls found at Zhoukoudian were broken. Another theory advanced by anthropologists was that Peking Man was a cannibal, who may not have eaten members of his own species as part of his regular diet, but who perhaps practiced a ritualistic killing and breaking open of the skull to eat the brain. This seems to have been Franz Weidenreich's contention, based on the preponderance of broken skulls found at the site, and the lack of complete skeletons.

TEILHARD SPENT THE FIRST FOUR MONTHS of 1930 writing his contributions to the monograph Davidson Black was editing about the *Sinanthropus* finds. He also worked on a number of his own papers on the geology of China. But he did not do this at home or in his office—he wrote while on the road. Teilhard took another field trip, this time north to Manchuria and then west to Mongolia, where he explored a number of paleontological sites. He returned briefly to Peking, then left on another trip, this one arranged by the American Museum of Natural History, to the Gobi Desert.

Teilhard, who had been accustomed to the austere travel arrangements of Father Licent, was in for a pleasant surprise on this expedition. He marveled at the comforts arranged by the organizers of this journey; the Americans were determined to maintain a high quality of life even while in the desert. When meat was in short supply, he later recounted, the expedition leader, Roy Chapman Andrews, would mount his jeep with a rifle, drive off, then return with a gazelle tied to the rear. With Licent, Teilhard had traveled on the uncomfortable back of a mule; the American expedition had fifty camels, in addition to jeeps and trucks. With Licent, he would freeze in glacial winds; the

Americans gave him a fur-lined coat to keep him warm by day, and a plush sheepskin sleeping bag to retire to comfortably at night.

On June 29, 1930, after a month in the Gobi, the expedition arrived at a dried-up, 2.5 million-year-old desert lake, set up what they named "Wolf Camp," and began to dig. Within a few hours they uncovered the bones of fifteen shovel-toothed mastodons. A little farther away, in a geological formation estimated to be 36 million years old, the team uncovered additional fossilized remains—giant pigs, a claw-footed animal resembling both a horse and a rhinoceros, and other ancient mammals. In total, the expedition shipped ninety cases of fossils from the site to the United States. Some of these can still be seen at the American Museum of Natural History.

On August 2, back from Mongolia, Teilhard wrote to Marcellin Boule:

> On returning to Peking, I had the pleasant surprise of finding at Black's laboratory *a second skull of Sinanthropus*, identical to the first by form and also (fortunately) by its state of conservation. In this second sample, one discerns the beginning of the nasal bones, and some further details. . . . Black has made some casts (very good ones) of all the isolated pieces. Two weeks from now, he should be able to give an estimate of the cranial capacity, taken from one absolutely perfect piece as preparation.—As for what concerns your idea of the identity of *Sinanthropus* and *Pithecanthropus*, it agrees with one of my first impressions, which I had communicated to you. But Black is absolutely affirmative on the existence of an appreciable difference in the development of the brain. And now we have *two* skulls exhibiting exactly the same differentiating character with respect to the *Pithecanthropus*. This reduces the chance of an individual variation.

These two important aspects would remain key to understanding who Peking Man was. Within the time frame given by Teilhard, Davidson Black had indeed been able to estimate the cranial capacity of Peking Man, which he concluded was about 900 cc. As one modern scholar has put it, Peking Man was not terribly bright, compared with modern humans (who have a cranial capacity of about 1,400 cc), but he wasn't terribly stupid, either. A brain size somewhere between that of a human and that of an ape further confirmed Peking Man's place as a missing link.

As for the other point raised by Teilhard, the hypothesized identification of Peking Man with the other missing link, Dubois's *Pithecanthropus*, this debate would only heat up in the following years. Both Dubois and Black would assert the independence of their fossils from each other. Most scientists, however, would argue that the Chinese and Javanese hominids represented two subpopulations of the same species—one we today call *Homo erectus*.

And when similar fossils were discovered in Africa later in the twentieth century, this debate would widen, with some scientists affirming that all three populations—the African, Javanese, and Chinese—belonged to the same species, and others using the designation *Homo ergaster* (Working Man) for the African *Homo erectus*, implying a different species. This distinction would be based on slight differences between the skulls found in Asia and skulls from Africa. The advocates of *Homo ergaster* believe that modern humans and Neanderthals descend from this species (through the intermediate species *Homo heidelbergensis*, a younger hominid found both in Africa and in Europe). Other experts believe that humans descend from one species, *Homo erectus*, which included Peking Man as one of its populations.

IN EARLY SEPTEMBER, Teilhard was informed that his father was ill, and he immediately left for France. He visited his parents at Sarcenat, then made a trip to Paris. There he received the bad news that Jesuit officials in Rome had ordered a second reading of his *Milieu Divin*, with the intention of preventing its publication. In addition, the diocesan censor of Malines, in Belgium, had refused permission for the publication of Teilhard's scientific essay in the *Revue des Questions Scientifiques*. As was usual by now, the Jesuits were not interested in any compromise with Father Teilhard de Chardin.

Teilhard had assumed a greater role in science, and his ideas were being debated by scientists around the world and receiving attention among the general public. The Jesuits, accordingly, were more afraid that these ideas would undermine the position of the Church. A priest who embraced evolution was bad enough, but one who was a prominent scientist, making advances in the study of evolution, was even more dangerous to his church.

Chapter 12

THE YELLOW CRUISE AND
THE MONGOLIAN PRINCESS

While in Paris, Teilhard met Georges-Marie Haardt, a decorated World War I hero who was working for the automobile tycoon André Citroën. Haardt invited Teilhard to join Citroën's latest adventure, an expedition of Citroën Kégresse P17 all-terrain tracked vehicles, over an 8,000-mile trek between the Mediterranean and the China Sea. This was to be the Croisière Jaune, the Yellow Cruise.

Haardt had led two such expeditions for Citroën, one in Alaska, and one in Africa. In the latter, the Black Cruise, Citroën vehicles traversed the continent from Algeria to the southeast coast and crossed over to Madagascar. But the planned trans-Asian expedition would be Citroën's most ambitious and daring project yet. The vehicles and personnel would penetrate into large parts of the world that seemed impassable by automobile, crossing unbridged torrential rivers, ascending steep mountains, and journeying across the empty expanses of the Gobi Desert—freezing at night and scorched by day—and would navigate through poorly mapped areas, some embroiled in civil strife and plagued by lawlessness.

Ostensibly, the purpose of the Yellow Cruise was to show off the capabilities of the Citroën vehicles—they would go where no motor vehicle had ever gone. But the organizers had other aims. This expedition would be a symbolic reconnection of East with West, a long-dreamt reopening of the fabled Silk Road between Asia and Europe, explored by Marco Polo almost seven centuries earlier. And equally, this was to be a scientific mission in the spirit of the investigations of nature that had begun in the nineteenth century. Teilhard would be recruited for this mission because of his expertise in geology.

Teilhard accepted. He was eager to escape his woes in Europe, his father was doing better, and as always he was intrigued by what Paleolithic finds Asia might yield. Perhaps he would discover hominids similar to Peking Man on this trek.

In January 1931, Teilhard traveled for the first time to the United States. He was well known in academic circles in America, and was respected as a scientist. He used this opportunity to deliver two public lectures in New York, both of them in English. One was at the American Museum of Natural History, and the other at Columbia University. People knew about the discovery of Peking Man, and the lectures were well attended. In New York, Teilhard met a director of the Citroën expedition who happened to be in the city, and finalized his arrangements to participate in the Yellow Cruise, and then left for Chicago for some public lectures. Being accustomed to oceans and deserts, he found the scenery of Lake Michigan a refreshing change.

Teilhard went on by train to San Francisco, where he would board a ship for Asia; he regretted not detouring to visit the Grand Canyon. He was pleased with his American trip: Everywhere he went, he felt welcome, and the American way of life and American attitudes to science and to scientists agreed with him. In California, he visited Berkeley and met with professors and students of anthropology.

Teilhard crossed the Pacific aboard the *President Garfield*. The ship stopped in Hawaii, then continued to Japan, where he spent some time visiting the temples of Kyoto. He was satisfied to find "golden altars, without idols." Once he was back in China, he was ready for the expedition.

The Yellow Cruise was a major technological undertaking. It aimed to prove that the harshest and most treacherous terrain could be negotiated by the carmaker's new vehicles. The people handpicked to take part in this historic attempt to cross Asia were divided into two groups, which were to move toward each other. One contingent would travel east from Beirut, the other west from Tientsin in the direction of the Pamir passes (in the Hindu Kush Mountains on China's western border). Teilhard was a member of the team traveling from Tientsin. At an arranged location, the oasis town of Aksu in the desert of western China, the two teams were to meet and continue together to Peking.

The Citroën half-tracks were equipped with treads specially designed for travel on roadless terrain, and Teilhard's group had seven of them. The group missed its scheduled departure date because one of the vehicles had a faulty tread, but the problem was resolved before the morning of May 12, 1931, when the convoy rolled west.

Ten days after their departure, Teilhard wrote his cousin Marguerite:

Already made 800 kilometers since May 12 across beautiful Mongolia. Second day, snowstorm; third day, sandstorm. Now, beautiful weather. Already, in these few days, I've gathered a number of important geological facts that are truly "illuminating" [in English in original, as is "captivating" below] on the structure of the Gobi. It's encouraging for the future. The kind of rapid research I've envisioned proves itself possible and fruitful. Pleasant journey, otherwise. This immense expanse, like the sea, crisscrossed by rocky ridges, populated by gazelles, sprinkled with white and red lamaseries, traversed by Mongols in modest clothing, is extremely "captivating."

The expedition gave Teilhard a deeper understanding of the geology of China. He also had much time to think. As the team traveled deeper into Central Asia, he continued to develop his evolutionary theory. Surrounded by the breathtaking scenery of stark, steep mountains, deep ravines, and ancient riverbeds, he thought about God and how his faith fit with his vision of evolution.

To show their acceptance of a priest among them, one late evening, in the mountainous Asian desert, the expedition members teased the Jesuit good-naturedly. When one of them presented a skit of a preacher, Teilhard retorted: "That was a Dominican." After the cruise was over, Georges-Marie Haardt would write to his wife: "Father Teilhard de Chardin is a Prince of the Church, but he has, as much as is possible, the spirit of the expedition."

The Europeans were allowed to travel through this territory only because they had Chinese colleagues with them. And Teilhard was the European in the group whom the Chinese trusted the most; whenever the team reached a regional frontier, he was the man they chose to sign the documents demanded by local authorities for permission to enter new territory.

On June 26, the expedition arrived at a dismantled fort: traces of a wall and four towers in ruins, with signs of burning. This was at a deserted junction of Central Asia, where the road between the Gobi Desert and the Tarim River Basin intersected with a road leading southeast from Sinkiang. They inspected the burned-down fort and, next to a well, saw a rock inscribed with Chinese characters. The translator, a man they called Petro, rushed over and read: "Do not go west. Danger. Hide your camels in the mountains and wait."

The thirty or so men congregated around this mysterious warning. They understood that the ominous message had been written recently—perhaps it was a warning by the head of an expedition to his men who were following. Roadside rocks were used as message boards in these remote, sparsely inhabited regions. "Grace of God . . ." someone muttered. But the team pressed on.

On June 28, after traveling some sixty miles from the junction and now in view of the snowy peaks of the Karlik Tagh, the team arrived at a village on the edge of the Hami oasis. The village had been burned and abandoned. A man evidently of Turkic origin, in rags, approached them in a hurry and said, in barely intelligible words: "Do not proceed. Do not go west. There is fight . . ." They tried to understand. "Who is fighting?" their translator asked. The man answered, "Everyone." Turkestan was embroiled in a bitter civil war.

But they continued west. The next day they encountered heavy fighting between local government forces and rebels. They saw many dead and dying, and Teilhard, with his war experience as a medic, helped the expedition physician, Dr. Delastre, dress the wounded.

As they proceeded through the desert plateaus, Teilhard wrote: "The greatest recompense of research lies in arriving at the point where reality appears in its deepest and most virginal state." He was approaching that state of nature as the expedition descended into the Turpan Basin, at its lowest point five hundred feet below sea level, where the Tarim River streams into Lop Nor. Teilhard saw this lake area as "one of the most sacred and mysterious" geological regions of Asia.

This primeval valley is surrounded by towering ancient mountains, the Tien Shan. Astute scientist that he was, Teilhard knew that the geological formations he observed here held sediments that might reveal fossils of an extremely old age. It might have been priceless to dig here. But there was no possibility for stopping: The region was dangerous, its sparse population was violent, and the team had to reach the arranged meeting point.

Soon they ran into new difficulties. The brigand governor of Sinkiang, Marshal King Cheu-yen, had the members of the team summarily arrested. An agreement had been made with him before the expedition began, which was to guarantee them safe passage through his territory despite its being closed to Westerners. King now was making demands—essentially, for ransom. "The local authorities, who lack many things, have facilitated your passage so far in this country that is so poor and with difficult roads," he told Teilhard and his companions. They understood their situation.

The head of the expedition radioed Paris. The captors explained that only if they were given radio equipment and Citroën vehicles would the group be allowed to continue west. From Paris, Citroën management promised that vehicles and radios would be sent to the governor through Moscow. The travelers were given permission to go as far as Ürümqi, the capital of Sinkiang, where they were kept in semi-imprisonment while their captors awaited the ransom. Assisted by the translator Petro, Teilhard—with his cultural sensitivity and personal charm—did most of the negotiations with the local leader.

From Ürümqi, Teilhard wrote Marguerite:

August 27, 1931

We are settled inside a pagoda that dominates Urumqi, surrounded by tall trees. Admirable view: interminable chain of the Tian-Shan, with summits of 5,000 to 6,000 meters, the snow shimmering. A few days ago, I walked in a forest of fir trees, which climbs up to 2,000 or 3,000 meters. One would think these are the Alps, if the chalets had not been replaced by Kirghiz yurts. Carpets of geraniums, of monkshood, of arnica. But these excursions are, unfortunately, an exception. For the most part, we are rather bored here. At age fifty, I find it vexing to lose time, while much work is awaiting me in Peking. But wasting much time is, of universal experience, one of the costs to consider when one adventures in Sinkiang.

Throughout their monthlong confinement, the team enjoyed elaborate lunches, and dinners followed by entertainment provided by exotically robed Mongol dancers. Here Teilhard made the acquaintance of Nirgidma of Torhout, a lady of the Mongolian nobility, also known as Princess Palta. She invited him along with a few other members of the Citroën group to her regal yurt, which was furnished with fine furs and red and black lacquerware. Her bar was stocked with the finest Alsatian kirsch and the most expensive Cognac.

Nirgidma was always perfumed and coiffed in Parisian style. Teilhard could hardly believe that this aristocratic Francophile with long red fingernails, who spoke perfect French and recounted her adventures flying with Antoine de Saint-Exupéry and learning to sing from Isadora Duncan, lived isolated like this in the middle of the Central Asian steppes. Earlier in her life, he learned, she had traveled extensively in the West and lived in Paris, absorbing much of the French culture she so admired. Some members of the expedition recalled seeing her at society events in Paris.

The princess enjoyed her conversations with the Jesuit, discussing life and the place of God in the universe. He was taken with his new noble disciple; some years later, Lucile Swan would sculpt a statue of Nirgidma. But as their confinement continued, the men lost patience and wanted to speed the process of their release. They decided to send a radio message to André Citroën to ask about the ransom. Yet how could they do this without arousing suspicion?

The group planned a fake celebration as cover. The Frenchmen invented a national day honoring the Third Republic, with flags and music. The cord attached to the flagpole flying the tricolor hid the antenna of their high-powered radio transmitter. In order to distract the Chinese soldiers guarding them and mask the sounds of the Morse key and the generator supplying power to the transmitter, the team brought out a phonograph to play military tunes in honor of the holiday.

Teilhard, who was responsible for operating the phonograph, selected a record he thought contained marching tunes. The Frenchmen stood at attention in expectation of the military music. The Chinese soldiers imitated their stance. When the record began to play, they heard a woman's voice, singing, *"Parlez-moi d'amour . . ."* The

Frenchmen didn't blink, and the Chinese facing them imitated their lack of response. The radio message went through, concealed behind the love song emanating from the phonograph.

In early September, the ransom finally arrived: three sedans and several crates of radio equipment, all sent by André Citroën from Paris. The expedition members were now released, and they split into two groups. One left for Peking, and the other, which included Teilhard, resumed the journey west to meet the team coming from Beirut. As planned, they met at Aksu, in the shadow of the Tien Shan Mountains, on October 8, 1931. André Sauvage, a member of the eastbound group, remembered meeting Teilhard, who was the first man to greet the other team. He thought Teilhard was

> capable of walking on water . . . A man of unequalled style; of a self-effacing and irresistible distinction. His voice, his diction, which had the tone of a harpsichord, his smile, which never quite turned to laughter, impressed themselves on anyone who was in the least attentive. Total lack of ecclesiasticism. As simple in his gestures as in his manners, but with the simplicity of a stele. Anxious to welcome, but like a rock of marble.

The combined team now drove toward Peking. On the way, Teilhard was quick to spot prehistoric stone tools lying in the sand and collect them for his lab. Every few miles he would order his driver to stop when he spotted reddish stone tools against the sandy background. These were the remains of an early hominid culture that expanded its domain from Siberia, Teilhard hypothesized. At Bäzäklik, near Turpan, east of Ürümqi, Teilhard took out his geologist's hammer and broke some rocks atop a ridge. He was exploring evidence linking the geological structure of the Gobi with that of Sinkiang, which would give him a clearer picture of the geological unity of central Asia.

On New Year's Day 1932, the members of the Yellow Cruise

Bäzäklik, some twenty-five miles from Turpan in
northwestern China, a stop on the Yellow Cruise, 1931.
Fondation Teilhard de Chardin, Paris.

emerged from the shadow of the Tien Shan Mountains. At nine a.m.,
the group found itself by a mission, and Father Teilhard led them into
the church, where he said Mass. Men in sturdy boots and heavy winter
coats with fur collars stood around the priest, surrounded by Christian
images illuminated by Chinese lanterns. Teilhard began:

My dear friends, we are together this morning in this little church in the
heart of China, to begin, in front of God, the new year. God, for each
one here, certainly does not have the same precise image. But since we
are all men, we cannot escape, not one of us, the sentiment and the idea

that, above and beyond us, a supreme energy exists, one that we must well recognize—because it is superior to us—the magnified equivalent of our intelligence and our will. It is in front of this powerful Presence that we must collect ourselves a moment at the start of this year. Of this universal Presence, which envelops us all, we first request to reunite us, as in a common living center, with those whom we love and who are starting, though far away from us, this new year. . . .

The universal nature of Teilhard's message was not lost on his fellow travelers: He was one of them. They were on their way home now, but their adventures had not yet come to an end. On January 30, Teilhard wrote to Marguerite:

We have arrived at Pao-Téou, the terminus of the [railroad] line from Peking, and the end of our adventures and misadventures. . . . Two days before our arrival here, while passing through a small village, a band of semi-bandit soldiers brusquely opened violent fire on our convoy, which was somewhat dispersed at that moment. The vehicle just in front of mine took some fifteen bullets, above all in its trailer; on mine, they did not fire. We returned fire, and so strongly that a white flag immediately appeared in the hands of the "enemy," which declared that it had made an error. Nobody was touched, fortunately, and it only served to get a lot of attention to our expedition.

In a commentary about Teilhard's letter and this incident, Marguerite wrote that her cousin usually underplayed the dangers he faced in his travels. She quoted another source on the Yellow Cruise (a book by Georges Le Fèvre), which related that in this village the men encountered a hardened group of bandits dressed as soldiers, who commonly marauded around the left bank of the Yellow River between the Alashan and Ordos deserts.

The ambush was professional. Several highwaymen, pretending to staff an official military checkpoint, attempted to stop the caravan. The expedition leader, suspecting a trap, ordered the team not to stop. As soon as the caravan passed the checkpoint, other men came out of hiding around them and began to shoot violently. When the team vigorously returned fire, a white flag appeared, apologies about a mistake were given, and the leader of the bandits offered tea and his superior's visiting card. The card, in Chinese, read: "General of Independent Cavalry."

At noon on February 12, to great applause, the Citroën team motored into the grounds of the French legation in the foreign quarter of Peking. Teilhard was happy to be back. He wrote one of his brothers in France that the main advantage of this journey was the firsthand knowledge of the continent it had granted him. "Finally, I have close to doubled my understanding of Asia. Ten months of my life, even at the age of fifty, are not too high a price to pay for this."

Soon after the group's arrival in Peking, Teilhard received the news that his eighty-six-year-old father had died on February 11, the day before their arrival. Teilhard was filled with sadness. His father had given him so many things: "certain precise aspirations, without a doubt, and even more, a certain deep equilibrium upon which everything is built."

Teilhard was called by the China Geological Survey to make another trip, this one to Shanxi Province, with his Chinese colleague Yang Zhongjian. In September, after this expedition, Teilhard boarded the *Porthos* to return to France. He stayed there four months, and while there received more unwelcome news. Writings by others, such as Édouard Le Roy, that were similar to Teilhard's were being placed on the Index of Forbidden Books; after the banning of his book, Le Roy recanted his words. And there was a new provincial father in Lyon, who did not view Teilhard favorably.

Disappointed, Teilhard spent his time in Paris working on matters other than religion. He delivered two lectures to his Jesuit brothers and others about the discovery of Peking Man. In February 1933, he left

France aboard the *Aramis*; he stopped in Singapore and Saigon, where he gave a lecture on paleontology. He arrived in Peking to find a China under attack from Japan and in turmoil. The atmosphere had changed for the worse, and for security reasons, all scientific material of the China Geological Survey had been transferred to Peking Union Medical College.

Teilhard returned to explore geological formations in Shanxi Province. His findings bolstered his earlier conclusions about the geology of China; he now saw a continuity of structures across Asia, and that these structures resembled geological formations he had studied in Europe. On June 22, Teilhard left China again—this time for the United States, accompanied by Davidson Black. They were headed to Washington, D.C., for the Sixteenth International Geological Congress to be held in July. There were friends on board ship with them, and Teilhard had a full social calendar for the duration of the voyage, which kept him away from writing.

At Zhoukoudian, however, there was great excitement. A new cave, called the upper cave, had been explored, near the top of Dragon Bone Hill. This excavation yielded, among other treasures: a cranium in excellent condition, pierced seashells—perhaps used for jewelry— and stone tools. All of these belonged to a much later period than Peking Man—as recently as 10,000 years ago. The researchers also found remains of ostrich and elephant bones. Teilhard spent time analyzing these new finds. Yet his work offered him little consolation; he was greatly distressed by the renewed troubles with his order. In exasperated letters to friends he continued to complain that the Church was in a static state while the world around it was in motion. For many months he had difficulty sleeping.

ANALYSIS OF THE PEKING MAN FOSSILS from Dragon Bone Hill led Teilhard and Black to infer that Peking Man had mastered the use of fire. Teilhard studied the residue in the dirt of the Zhoukoudian cave floor and walls, and concluded that the carbon deposits there were caused by fire. Later chemical analyses would question some of his assumptions, and some researchers have posited that the black matter in the cave was not carbon but manganese. However, Teilhard found other evidence that Peking Man could make fire; he analyzed the fossilized animal bones found with Peking Man, and these showed signs of having been burned. His conclusion was that Peking Man was the earliest of our ancestors to use and control fire. Two research articles appearing seven decades later have corroborated Teilhard's analysis.

In July 1933, Teilhard and Black presented their findings about Peking Man at the Washington conference and discussed what was known so far about this fascinating hominid. Their presentation was anticipated with great excitement; there had even been a rumor that Black would arrive with a tooth of Peking Man. Black by now had a near-mythical reputation, and Teilhard's fame continued to rise. Zhoukoudian was an isolated location, without nearby hominid sites, and the conference participants resolved to search for comparable sites in China. The Congress commissioned an exploration, for the following year, of several areas of China, as far west as Sichuan Province.

When the congress was over, Teilhard participated in an International Geological Congress program, a train expedition across the United States. This was an unusual trip. The train moved only at night, and during the day it stayed in one location—a different one each day—and the participating paleontologists and geologists would spend the day exploring the terrain and looking for fossils. The expedition followed the Oregon Trail from Missouri across the country, and ended at Crater Lake, Oregon. One train car served as a conference hall, furnished with tables and chairs, blackboards, a film projector, and a screen. There was even a laboratory facility, in another car, for analyzing samples. Other cars served as sleeping quarters. These, however, were of two kinds: Married couples were given spacious and comfortable quarters, while participants without spouses shared more cramped compartments with others of the same gender.

One evening, after a long, tiring day spent digging in the plains of Kansas, Teilhard got ready for bed. He looked around for a place to hang his clothes, then shot a sad glance toward the front of the train, where the married couples enjoyed greater luxury. He looked at his compartment-mate, the Belgian scientist Armand Renier, and said, "Our colleague Provust, who brought his wife along, got the benefit of a special compartment. Why don't you use your diplomatic skills and get us a wife for two?" He winked at Renier, nodding in the direction of the next compartment, where a famous woman paleobotanist of a certain age was sleeping alone.

At the end of the trip, Teilhard went to California, where he had a marvelous stay. He had enjoyed visiting the United States before, but this trip was especially rewarding. He found the geology of southern California interesting in its variety, and he loved the unspoiled natural scenery around him: sun-kissed mountains, rolling hills, and green oaks yielding to cactus and yucca in the desert. "Life is so simple here, *unconventional*," he wrote. "One sleeps under the stars, and eats at any hour of the day or night in a bar, perched high on a stool. No one bothers you if you want tranquillity. After a month here, I feel at home in California."

But China beckoned. He was eager to resume his work on Peking Man with Davidson Black and other colleagues.

Chapter 13

LUCILE SWAN RECONSTRUCTS
PEKING MAN

In late 1933, Teilhard was again in China, and he soon was back at Zhoukoudian. The excavators had turned their full attention to the upper cave, where Late Stone Age artifacts and fossils were being discovered—those belonging to hominids that lived hundreds of thousands of years after Peking Man. These were anatomically modern humans who lived here as late as the end of the Ice Age. On the last day of 1933, Teilhard wrote the Abbé Breuil in Paris:

> I have returned to Choukoutien with Black fifteen days ago for the annual closure of the digging season. The clearing of the collapsed roof has finally begun (Pei [Wenzhong] has installed a cable-and-pulley to lift and remove the debris), and we are well on our way to entering "Cultural Zone A" come next spring. The "Upper Cave" is now exhausted, save for a small corner that has yet to be explored. It is a small chalky chamber, about twelve meters high (that is, completely independent of the resting place of the *Sinanthropus*), covered with stalagmites and stalactites. Its general aspect is very different from that of the "cavern" of the *Sinanthropus*.

In his laboratory at Peking Union Medical College, Davidson Black was overworking himself to the point of a breakdown. He had a congenital heart defect, and in early 1934 he showed signs of a serious cardiac problem and was hospitalized. Doctors advised him to rest. But Black told them that he simply couldn't do it, and soon he was back at his lab. He felt he could not slow down, because he had been achieving so much and there was more work ahead. February was a very productive month for Black and his team. They analyzed skulls from the upper cave, and gleaned a wealth of information.

On February 15, Teilhard apprised Marcellin Boule in Paris of the situation:

> Black has just finished making cranial diagrams of the "Upper Cave Man." One of the skulls is absolutely perfect. The other two are pass- ably damaged.—but with many usable parts. Black is careful not to prejudice his conclusions.—My impression is that all this will turn out to be something curiously similar to Cro-Magnon.—We are preparing a "Preliminary Report." I've told you that among the rest of the animals found in the Upper Cave there were remnants of an ostrich (two femurs and vertebrae) whose existence until now had only been known through its eggs.

On the night of March 15–16, Black was working in his laboratory. At around seven p.m., his assistant, Dr. Paul Stevenson, came in and spoke with him for about half an hour, then left. When Stevenson returned to the lab later that night, he found Black lying on the floor, his desk covered with papers and many fossils from Zhouk-oudian. Stevenson called for help, but it was obvious that Black was dead. He was just forty-nine years old. His body was transferred to the "cold room" in the medical college, and arrangements were made for burial.

Teilhard, who heard of this tragic loss from the staff at the college on the following day, was inconsolable. He wrote Lucile Swan:

Lucile, dear friend,

I send to you these few lines, à tout hazard, in the hope that they will reach you in the south. Yesterday, coming back to Peiping [Peking], I found with joy your precious letter of the 12th. I received it in my heart and in my mind. Yes, I feel that they are great prospects for us. . . . If I can give you something, you, in turn, you can help me and complete me, not only by the warm light of yourself, but also by your keen and strong sense of reality. You have still to teach me a lot of things. And may be we shall do something for "dematerializing" a little the Matter around us.

I feel deeply, today, this necessity for saving the world from its material darkness. You know already that Dr. Black passed away, day before yesterday. The apparent absurdity of this premature end,—the stoical, but blind, acceptance of this fate by the surrounding friends,—the complete absence of "light" on the poor body lying in the ice-room of the P.U.M.C. [the medical college], have deepened my grief, and revolted my mind.—Either there is an escape, some-where, for the thought and the personality,—or the world is a tremendous mistake. And we must stop.—But, because nobody will admit that we must stop,—then, we must *believe*. To awake this belief must be, more than ever, my duty. I have sworn it to myself, on the remains of Davy,—more than a brother, for me.

The end was nice, and simple. Black had the fallacious impression of getting much better. He had come to the Laboratory, seen several friends, and talked cheerfully. Then, when he was alone, he passed away, close to his desk, between his maps and his fossils. Absurd,—or wonderful.

Now we try to save the boat. I have, for a part, to take the rudder. Dr. Greene [Roger Greene, the executive head of the medical college] asked me to do so.—The first clear thing is that we have to go on. The plans will not be altered. Most probably I will go to Shanghai (maybe arriving on the 24 only), and we shall try to make the trip Nanking–Hankow.—In the meantime, we shall search somewhere in the world an anthropologist for the study of our material. Every day, probably, the depth of the loss we have made will become more apparent.

Teilhard needed help at Zhoukoudian, and he arranged that a telegram be sent to his friend George Barbour, a geologist with the Zhoukoudian team who was then on a boat near Hilo, having lectured about volcanoes in Hawaii. At three a.m. on March 17, a radio telegram was brought to Barbour's cabin, and he was awakened to read: "Davidson Black died heart yesterday."

Important changes would be necessary at Zhoukoudian, Barbour knew, and he and Teilhard were now responsible for the continuity of the operation. Barbour boarded a ship for China at once. Ten days later, as the *Empress* reached Shanghai and anchored off the Bund, a courier came aboard and handed Barbour a calling card with the scribbled message: "I wait for you at the Custom's jetty. Teilhard." Barbour hurried ashore to meet Teilhard, then went with him to a nearby hotel for lunch. They spent two hours reminiscing about Black, and Teilhard described their colleague's last days. "He was like a brother to me," he kept repeating. They talked about what had to be done now.

Teilhard took charge of the Peking Man research project for a time, but because he was involved in other work, the team hired Franz Weidenreich as a permanent replacement for Davidson Black. The Jewish Weidenreich had been fired from his full professorship at the University of Frankfurt am Main in Hitler's Germany and had eventually found a visiting position at the University of Chicago. His efforts on the Peking Man project would prove invaluable, and he would continue his work beyond World War II, when he took a visiting position at the American Museum of Natural History.

Weidenreich was stepping into big shoes, and he knew it. Black had

been an exceptionally well loved director of the Peking Man enterprise, and Weidenreich didn't have his accessible style and easy rapport
with those who worked for him. At first he was even suspicious of the
way the Chinese workers were handling the fossils. He noticed that
skulls alone seemed to be discovered, and he feared that the workers
were discarding fossils of other bones. Only after observing their
behavior on site for some time was he reassured.

Weidenreich's management style was much more formal than
Black's; he would visit the site at Zhoukoudian in a three-piece suit. He
was fastidious, and tight with money, apparently not understanding
how foundation money is generally administered—for example, he
didn't quite understand that funds that were not used were lost, and
thus he conserved funds overzealously, and needlessly. Yet he was a
superb anatomist, and he so brilliantly described the Peking Man finds
that, decades later, we know much about *Sinanthropus* because of Weidenreich's carefully written descriptions.

WEIDENREICH'S PLAN WAS to synthesize all the details he and others had been uncovering about Peking Man. He wanted to combine
these pieces of information to create a flesh-and-blood image of what
Sinanthropus actually looked like. For this he needed someone who
understood both anatomy and sculpture, someone who was nearby
and whom he could guide. In Peking, there was one person who was
all these things: the American sculptress Lucile Swan.

Weidenreich hired Swan full-time to work on re-creating Peking
Man. She would use everything science had learned about this hominid
to fashion a likeness. Every weekday she went to Weidenreich's lab
at Peking Union Medical College, where she worked closely with him
and his assistant, a young German woman named Claire (Hirschberg)

Monumental bust of Peking Man, based on the work of Lucile Swan,
at the Zhoukoudian site. Photograph by the author.

Taschdjian, who was responsible for the fossils kept in the lab. Swan
gathered data, took measurements, read reports about the *Sinanthropus*,
and made casts of fossils. Then she took all the information, as well as
the casts, to her studio, where she worked on the statues of the
hominid. This was a matter of both science and art. Over a few months,
Swan completed several statues, believed then as now to be excellent
likenesses of Peking Man.

Swan's employment with Weidenreich and the frequent visits she
made to the medical college put her in closer contact with Teilhard.
She saw him not only in the afternoons at tea, and often for dinner, but
also throughout the workday. Their friendship grew stronger and
deeper with the increased contact.

Teilhard, frustrated with his treatment by the Church, seeing his
requests for permission to publish books forever turned down, and
having his pleas for leaves to return to France often fall on deaf ears,
needed a friend. He was accustomed to female company, enjoying
close relationships with his sisters and cousins, and he thought that
Lucile might play a similar role in China.

Lucile Swan was sensual and flirtatious, vivacious and interesting, and men were drawn to her. Around the time she met Teilhard, she was being pursued by a number of eligible men within the Peking expatriate community. But soon after her relationship with the priest began to develop, she abandoned her suitors and devoted all her time and attention to him.

When she started to work for Weidenreich, Swan and Teilhard were seeing each other every day—unless he was away traveling. And while he had friends around the world, Lucile had fewer close friends, and therefore greater need of him than he did of her. The two would spend hours in intimate conversation. They often went out for picnics together, and traveled to the Western Hills outside the city, where they enjoyed the scenery and conversation. For Lucile, Teilhard was her closest friend and confidant, a sensitive individual with whom she could share her deepest thoughts, hopes, desires.

Lucile was falling in love. And in Teilhard's own self-protecting way, he was in love with her. He started addressing her in his letters as "Dear friend," then progressed to "Lucile, dear" and to "Dearest Lucile." Finally he called her simply "Dearest."

But there was a glaring inequality in the relationship. While he viewed her much the way he viewed a close sister or cousin, the platonic barrier of such relationships did not exist in the way she viewed him. Lucile was an adult woman, with the normal needs of a woman. She had been married before, and naturally she expected a relationship in which someone could give her the closeness she had learned to expect with a man. She wanted sex.

Teilhard recognized this, but he was not prepared to break his vow of celibacy for her. Perhaps he even felt that he had already broken some form of trust that his church had placed in him, and therefore was more determined not to give up his chastity. Yet although he realized the crucial disparity between their needs, Teilhard did not push her away. He needed her friendship too much.

Lucile did not want to give up on Teilhard, either, even if he was a priest who insisted he would never break his vows. She kept pressing him for the intimacy she desired—gently, at first, and later with greater insistence. While she understood that he was a priest, or at least said she understood, her normal human instincts and desires were working against that rational understanding. And as the two were spending much time together, it was hard for her to contain her desire for this handsome man, even when he did wear a cassock (as he did only occasionally). Whenever he suggested that she make friends with other men, perhaps get married, she would answer that she couldn't. As their relationship progressed, it became increasingly complicated.

ONCE FRANZ WEIDENREICH was settled in his role as director of the Peking Man project, Teilhard was released from many of the responsibilities he had assumed after Davidson Black died. The Rockefeller Foundation had given him great latitude in realizing many of Black's plans. Among these was the involvement of George Barbour in special projects at the site. Since they both had more free time with Weidenreich in charge, Teilhard and Barbour decided to conduct joint research.

They embarked on an expedition far inland, to Kweiling (Guilin) and beyond, in picturesque Kwangsi (Guangxi) Province in southern China. They were studying the geology of the country, and comparing it across regions. They were also looking for evidence of early human habitation. At one point, Barbour caught Teilhard on film, bathing nude in a mountain stream. When he saw the photograph sometime later, Teilhard commented, "Do you think that if the head of my Order in Rome saw me like that in the middle of the river he would consider me prematurely unfrocked?"

From the road, Teilhard wrote Lucile:

Kweiling (Kwangsi), January 28, 1935

Lucile, dear

My last letter was from Nanning—close to Indochina. Now we are almost in the north of the Kwangsi, not very far from Hunan. This is the extreme point we planned to reach. . . . Each day brings an opportunity for new observations which seem to confirm our views concerning the Tertiary and Quaternary Geology of S. China.—In prehistorical matter, we did not find any Choukoutien. But, here as in Nanning, we have excellent evidences of an ancient culture (cave-dwellers, shell-eating people) which marks an interesting stage in the prehistory of China. Unfortunately, all the caves have been "devastated" by fossil-hunter Chinese, so that a few patches only of the deposits are preserved.

He promised to write her again from Hong Kong, from where he and Barbour were to sail to Peking at the end of their expedition. He ended that letter, datelined Hong Kong, February 8, 1935, with the following: "A bientôt, dear. / And believe me / Yours very deeply / Pierre."

In Peking, the priest continued his routine of visiting his sculptress friend every day for tea. They picnicked with other friends, members of the social group around them. These outings often took place under the trees in the park of the majestic Temple of Heaven, a complex of three circular sacred buildings begun in 1420 for the emperors of the Ming dynasty, within a large wooded park. Lucile helped Teilhard with his work: she typed many of his papers, and at times translated them into English. In his letters, he called her his "compass point" or his "light." Physical desire, though, was a key element of her feelings for him—and something that was becoming very hard to control. On Valentine's Day she wrote in her journal:

I have been pretty wise and I have not been depressed—but my friend—the salt has gone out of life when you are away—You must help me to *see* a way—Friendship is no doubt the highest form of love—and also very difficult—my primitive woman instincts are so strong—to learn how to control this love is so difficult—but oh my beloved what a worthwhile line of effort—

The constant pressure to reciprocate her affection for him in a physical way compelled Teilhard to contemplate the meaning of sex. As a response to these urgings, and to clarify to himself the nature of their relationship, he wrote an essay titled "The Evolution of Chastity." It was the manifesto of a man who was attracted to women but because of his vows could not experience physical love. The view of woman as expressed in his essay is that of Dante's Beatrice, who encourages a man to advance beyond himself toward God.

Lucile's deep frustration at not being able to consummate her relationship with Teilhard made her restless. At one point she abruptly left Peking to spend time alone in the countryside and ponder her predicament. Another time, when Teilhard returned to Peking from one of his trips, she confronted him with the news that she was leaving for an extended period in order to exhibit her sculpture in New York. On March 29, 1935, Teilhard wrote her:

Lucile, dear,

This time, *you* are leaving, and I am left behind. This is the hardest part. But I am glad it is my share: because, you know, I would like so much to take on me any pain from your life.—My dream and my hope, is to be for you a strength and a joy *only*, Lucile. . . .

You are going East. After a few days I am also leaving, West. Is that not as if, by our two lives, we were making the symbolic gesture of "encircling" the Earth?—Be *sure* of that: the separation of today is not the end of anything: just a start for the new life.—

God bless you, precious

Pierre

And just at this time, the Abbé Breuil came to visit Teilhard and Weidenreich. Teilhard was busy taking Breuil around Peking, visiting Zhoukoudian, and showing him the paleontological finds of Peking Man. In early April he wrote Lucile:

Since I left you, on the Tientsin platform, eight days have already passed.——We were only two days in Tientsin, Breuil and myself. On Sunday night we were back in Peitang [in Peking]. And since then the routine has been going on——except that I feel still a little lost at 5 p.m. [Teilhard and Lucile's usual teatime]. . . .

God bless you, dear. And might the big ocean be always smooth and bright for you.

Yours so much

Pierre

Today (Sunday morning) we are going to the Ming Tombs (Breuil, Pei & Myself). I will think of you at each blossom of tree, I think.

Seeing that Weidenreich had assumed his duties so confidently, Teilhard felt comfortable to travel. He took the Trans-Siberian Railway from Vladivostok to Moscow together with Breuil, and then returned with him to Paris.

Teilhard went on to England for a meeting of the British Geological Survey. He returned to Paris and then traveled to Malta, Bombay, and Kashmir, where he took part in a geological expedition. According to Madeleine Barthélemy-Madaule, who has studied his life, in the years 1923 to 1946, Teilhard never remained at any one location continuously for more than fifteen days.

Even with all this movement, he kept up with his many friends, including Monfreid and the Mongolian princess Nirgidma, through letters and visits. On June 16, Teilhard wrote to Lucile from Paris:

A fortnight ago, I had lunch à Neuilly with the "Pirate" [Monfreid] and his most respectable family. In such occurrences, I become myself a member of the family, and I appreciate just so much as before this warm friendship. Monfreid hopes to integrate Abyssinia with the Italian troops; and for complex (personal and political) reasons, he speaks, writes, and lectures as a convinced supporter of Mussolini.— War is expected to start there in August.

Nirgidma is on her way back from Palestina. She has been sent by newspapers to Morocco, Egypt, etc., in order to study the Islamic question (why she, a Mongol?)—and I heard that some of her articles have been already published. I expect to see her at the end of the month.

PEKING MAN WAS NOW FLESHED OUT in more detail as a hominid from half a million years ago, a member of a well-adapted species that lived in Asia for hundreds of thousands of years and left its mark in a cave southwest of Peking. The international team that discovered him was now occupied with explaining further to a world in search of knowledge about human origins the importance of this evolutionary link with the past. The next step for anthropology was to locate Peking Man more firmly within our evolution from our common roots with the apes.

In Africa, in the 1930s, Louis S. B. Leakey (1903–1972) had begun to dig in Olduvai Gorge. An African-born paleontologist of missionary parents, Leakey was convinced that the most ancient origins of humanity lay in Africa, not in Asia, despite the earlier discoveries in Java and China.

Over several decades, Leakey, his two successive wives, his sons, and his daughter-in-law would discover the next link in our human ancestry chain. Indeed, their finds would constitute the earliest

Teilhard de Chardin (left) and the Abbé Breuil visiting the Ming dynasty (1368–1644) tombs north of Peking on April 7, 1935. Photograph by Pei Wenzhong, discoverer of the first skull of Peking Man. Teilhard de Chardin collections, Georgetown University.

members of our genus (the australopithecines are more primitive and not considered members of the genus *Homo*). The Leakeys found partial skeletons and skulls of hominids in Africa dated 2.2 to 1.6 million years ago, and Louis Leakey's "Out of Africa" theory was confirmed by analysis of fossils of these hominids, which he named *Homo habilis* (again, Able Man or Handy Man) because they used tools. With our ancestry's development now clearer, Peking Man could be seen as even more important, since he was indeed the first human ancestor known to make and use fire.

Our closest living relatives in the animal world are African: chimpanzees (including bonobos) and gorillas. This genetic finding has been known only for several decades. Before, scientists believed that humans were equally close genetically to all apes—both African and Asian. Early in the twentieth century, many scientists believed that Asia may well have been our place of descent from the apes. But molecular evidence strongly favored an African origin. Humans have been shown to be genetically closer to the African chimpanzee than to Asian apes. Thus scientists began later in the century to look closely at Africa for a common link between ape and human.

Around 6 or 7 million years ago, the line leading to today's humans diverged from that of chimpanzees. The two very early hominids discovered in Africa at the start of the twenty-first century, *Sahelanthropus tchadensis* (Sahel Man of Chad, Sahel being the semidesert border of the Sahara extending from Mauritania to Chad) and *Orrorin tugenensis* (named for the Kenyan village of Tugen), date from near the time of the divergence of our ancestors from those of chimpanzees.

Chapter 14

PEKING MAN VANISHES

In the fall of 1935, Teilhard traveled to the Salt Range foothills of the Himalayas. He also explored Paleolithic sites in the Sohan River Valley in the state of Punjab in northern India. There he found cutting and scraping tools that he dated to around the time of Peking Man. In December, he went to Java. He was traveling with his friend Helmut de Terra, a Yale University geologist born in Germany, whom he had first met in Washington at the International Geological Congress of 1933. The two scientists visited the site along the Solo River where Eugène Dubois had discovered Java Man some four decades before.

The paleontologist G. H. R. von Koenigswald had been excavating this site further since the early 1930s, and when Teilhard and de Terra arrived, he accompanied them on a ten-day trek through the country-side. They slept in native huts along the way and recruited local people to help them dig for fossils. They visited a site that Koenigswald had been excavating on the Solo, which yielded fossils similar to those found at Zhoukoudian, and together they hypothesized that Java may have served as a crossroads of early hominid civilizations: one of tribes traveling south from China, and another of groups traveling southeast from India. We believe today that both Java Man and Peking Man, as well as hominids whose bones were found in Africa—most recently in

Nariokotome in Kenya in 1984 and at Olduvai Gorge in 1986—and in Soviet Georgia in 1991, are all *Homo erectus*.

The trip confirmed in Teilhard's mind that scientific work on evolution was an important goal in life, and it energized him to continue this work so that he could reconcile Christianity with the wonders of evolution. But while returning by ship to China, he was aware of the pressing problem of his relationship with Lucile. He and Lucile had now been apart for almost a year. The separation had underscored the problem of the future of their relationship. Lucile wanted greater intimacy. Teilhard, vulnerable because of the problems with his order, did not want to lose the most important friend he had.

In his letters to her, he addressed the problem of his celibacy. He wrote that he understood her objecting to his celibacy because she viewed it as a way of denying one of the most fundamental laws of the universe. But procreation, he pointed out, was not the only reason for sexual relations; sexuality was also a manifestation of love. (Here Teilhard's view diverged from the Catholic teaching that sex is only for procreation.) This form of love between a man and a woman, Teilhard explained, was simply not possible for him. He told Lucile that he did not want to delve into this issue too deeply. Perhaps he was trying to buy himself more time.

Any confrontation was avoided, at least temporarily, by the fact that more travel lay ahead for him. Upon his arrival in Peking in early 1936, Teilhard found a letter informing him that his mother had died at Sarcenat on February 7, and he headed back to France. A year later, having returned to China from France, he embarked on another voyage to the United States. He sailed from China on February 25, 1937, landed in Seattle, and went by train to the East Coast. He had come to lecture on Peking Man, which he did widely, and he brought a cast of the best skull from Zhoukoudian to give to the American Museum of Natural History in New York. In Philadelphia, where he attended the International Symposium on Early Man, sponsored by the Academy of

Natural Sciences, he stayed at the house of Helmut and Rhoda de Terra, and they accompanied him to his lectures.

Teilhard explained the role of *Sinanthropus* in evolutionary theory and presented his ideas about the moment in human development at which hominids became rational beings. He likened that moment to the instant when water reaches the boiling point. He had formulated these ideas just before the trip, but they were the culmination of thoughts he had had throughout his life, inspired by the cave art he had observed in France and Spain, which evinced the first occurrence of abstraction and symbolic thinking in the history of human development.

Teilhard's lectures received considerable attention, including an article in *The New York Times* on March 20 that described Teilhard's lecture at the International Symposium on Early Man. The article identified him as a Jesuit, and stated the essence of his talk and the idea of the common ancestry of humans and apes. Many people misunderstood his views to mean that humans were descended from apes. Teilhard, seeing that he had been misinterpreted and that his remarks were being sensationalized by the public, gave more newspaper interviews in an effort to correct the misapprehension. But the damage had been done. It was one thing to explain evolution as a branching process of change, through which species evolve; it was quite another to say that people "came from apes." And since American Catholics were at that time sensitive about the theory of evolution, Teilhard had lost ground.

In Massachusetts, Teilhard was scheduled to attend a ceremony at Boston College, a Jesuit school, where he was to receive an honorary degree. But after the negative publicity, school officials changed their minds. As soon as he arrived there, he was told that the degree and his invitation to speak had been rescinded. The archbishop of Boston, William Henry Cardinal O'Connell, who had previously attended ceremonies at the college, said that he would not appear if Teilhard was honored. Cardinal O'Connell had been offended by Teilhard's statements on evolution as reported in the press.

Soon afterward, it appears, the Jesuit authorities in Rome learned of the media attention in the United States, perhaps through an article critical of Teilhard published in the Vatican newspaper, *L'Osservatore Romano*. The Jesuit leaders were apoplectic. Teilhard went to France, and once he arrived he was chastised strongly by his provincial in Lyon, who relayed to him the order from Rome: He would have to stay in Paris until further notice. But before any new message could be sent to him, Teilhard was admitted to the Clinique Pasteur with severe fever and chills. He was later diagnosed with malaria.

Despite his illness, Teilhard tried to work in the hospital. His renewed troubles with the Jesuits had convinced him that he should stop making any public statements about evolution, even if they were within a purely scientific context. Instead, he should write his complete philosophy of evolution as cogently as possible and publish it. Perhaps this notion was naive—after all, he'd been denied the opportunity to publish before—but he thought that the Jesuits would accept a well-constructed argument for evolution and thereafter leave him alone.

Teilhard had so many visitors at the hospital that he was never able to work properly. Only weeks later, released from the hospital and on his way back to China, was he able to resume his work.

The China he returned to was roiling. On July 7, 1937, an incident of grave consequence took place on the Marco Polo Bridge, on the road between Peking and Zhoukoudian. Soldiers of the Imperial Japanese Army shot and killed a group of Chinese civilians. This incident ignited the Sino-Japanese War, between a militarily prepared Japan and a deeply divided and weak China. The excavations at Zhoukoudian were halted for two days, resumed briefly, and were then suspended when Japanese forces moved south and took control of the area around Zhoukoudian.

The invading Japanese arrested several workers at Dragon Bone Hill and interrogated them about the project. The Japanese general in charge of the operation seemed especially eager to see the Peking Man remains. In the capital, still under Chinese control, the research team locked the relics in a storeroom at Peking Union Medical College and weighed their options.

Teilhard continued to travel after his return to China, despite medical advice that he rest. In December 1937, he went to Burma with Helmut de Terra. There they discovered the remnants of a stone-tool industry, but no fossils. Teilhard stayed in Burma until early April 1938, then went to Java for another excavation with G. H. R. von Koenigswald. He returned to China in May.

In Peking, Teilhard saw Lucile again. They managed to put their problems behind them temporarily and took a trip together, visiting Japan with a group of women missionaries. On their return, because of increasing political hostility between Japan and China, the Japanese captain of the ship carrying them home abruptly turned around; when the ship docked in Kobe, there was no place for them to stay. The group finally found a boat to take them back to China, and after they landed, they took a train to Peking. Teilhard was relieved to be in the familiar surroundings of the capital, with "its dust, its blue sky, its apple blossom, its rickshaws, its dilapidation, and the continual cry of pigeons passing overhead, with a whistle in their tails." The political situation in China, however, was critical.

In June, Teilhard was still formulating his philosophy of man, physics, spirituality, and evolution. He was trying to derive a convincing amalgam of scientific considerations about the present and future of humanity. The result, Le Phénomène Humain—The Phenomenon of Man—would be his masterpiece. But it would not be published for many years.

The book incorporated work he had been doing most of his life. His basic tenet was that human beings were the key to understanding evolution. Le Phénomène Humain progressed through four stages: pre-life, life, thought, and survival. In the first part of the book, he gave a survey of physics, in which he described the physical laws of the universe.

He then presented his hypotheses about the origins of life. He explained the evolution of life from simple to more complex organisms, how life on Earth progressed from plants to animals, and from simple forms to fish, amphibians, vertebrates, and mammals, including humans.

With humans, there evolved the complex elements of thought, cognition, and awareness—everything that distinguishes us from other mammals. Here Teilhard offered his idea of the noosphere (from the Greek *noos*, or spirit), the realm above and containing the biosphere. The noosphere is the sphere of thought and ideas.

Finally, the book addressed the future: human survival. To Teilhard, the universe was a single organism. All created reality, he stipulated, had an inside and an outside level, and there were two kinds of energy in the universe: the measurable force of physics and a force residing in human thought. The future of evolution, according to Teilhard, was the Omega Point, the ultimate site of convergence through evolution. God and Christianity are rarely mentioned in the book—mostly at the end. This was principally a work about evolution, not religion, at least not the way the Church viewed it. It's no wonder that the Jesuits didn't like it much.

His study of cave art had taught Teilhard that even early *Homo sapiens* could think symbolically and create things, unlike animals. After millions of years of evolution, a conscious animal came to inhabit the planet, one that made tools, conquered fire (as already Peking Man had), evolved consciousness, and created art, language, and thought. The war Teilhard had seen taught him that collectivization of humans meant not only alliances but also conflict. The process of evolution after consciousness, then, was fraught with danger. This brought Teilhard to consider the idea of survival, and of the future of humanity and the planet.

Teilhard asked himself ecological and environmental questions. What happens when wars end? Do populations continue to grow at the prewar rate? What happens to the competition for resources? How do collections of people—nations, tribes, ethnic groups—respond to one another? Can the "psychic temperature" of humanity reach a critical barrier? And would the process of evolution that led to the creation of human thought and creativity lead us to disintegration? These disturbing thoughts about the final outcome of evolution occupied his mind and directed his writing during this difficult period in history.

Teilhard had obligations as well. He had to deliver lectures in New York on his scientific work, and in September 1938 he sailed to Seattle and continued by train eastward. After speaking in New York, he left for France, where he stayed into the first part of 1939 visiting friends and relatives, including Marguerite. In late May he wrote to Lucile, in the United States: "Fortunately, you are waiting for me in sunny California, dearest! Most probably . . . Marguerite (I told you often about her) will come with me as far as New York; she needs to see America and Washington, before writing a book on [Abraham] Lincoln."

Teilhard made it back to New York and lectured at the American Museum of Natural History. He went to Chicago for another talk, and then to Berkeley, where he spoke about the geology of Asia at a meeting of the Geological Society of America. He was finally reunited with Lucile, not in California but in Vancouver, and on August 5 they boarded the *Empress* bound for Shanghai. They had most of a month together on the crossing, and in this time she helped him formulate some of the ideas for his new book. On August 30, two days before Hitler invaded Poland and World War II began, they arrived in China.

Throughout the rest of 1939, Teilhard labored on *Le Phénomène Humain*. He was constantly revising, because he knew that his request to publish the book would face stiff opposition in Rome. As a Jesuit, he could not contact a publishing house and make arrangements to issue a book without permission from the order. He kept searching for a formula that would allow him to say what he needed to say and at the same time minimize the friction with the Vatican and the Jesuit authorities. Every day he met with Lucile, who was helping him edit and typing his drafts. They were spending so many hours together that the tension between them, and her dissatisfaction with their relationship, finally came to the surface.

Sometime in October, Lucile wrote Teilhard a long letter. She never gave it to him; it was found among her papers after she died. The letter read, in part:

> What happens to cause this deep feeling of depression and outbursts like yesterday? It is true that things have not changed, at least your attitude has not changed. It is just that I understand it better, perhaps not "understand" but at least I know more about it. And I *am* convinced that the root of the whole thing is that you really do live on a different, a higher, plane than do most of us—and I have always considered you as a regular man—superior, yes, but nevertheless with the same needs as other men. And now I don't believe that is true—I have thought that there was a certain aloofness or coldness about you which I would help by giving without reserve a deep warm love. But I wonder if you either want or understand it. You love, yes, but on a different plane. . . . You just don't experience jealousy or some of the other less admirable emotions and so cannot understand them. They are quite normal in the "average" person—but anyhow it is all mixed up because I can't keep up to your plane and ask for things that you do not want to give because you really don't understand them—and then that causes an inequality that is *ugly*. And then these things happen and then I feel like hell—and what's it all about anyway.

Apparently Lucile could not bring herself to give this letter—which continued for several more pages—to Teilhard. Perhaps writing it made her feel better, and their relationship maintained its status quo.

Since the Japanese first invaded China, and more so after the incident on the Marco Polo Bridge and the declaration of war, a sense of peril had been growing within the expatriate community in Peking.

With the outbreak of war in Europe, many foreigners left for other destinations in Asia and elsewhere, fearing that the Japanese occupation of China might become more brutal. As the expatriate community around them shrank, Teilhard and Lucile drew closer.

The Americans, among others, recognized the danger to U.S. interests in China and the possibility that the Rockefeller Foundation would or could not offer long-term protection to scientific interests in this land. There was a strong likelihood as well that the United States and Japan would soon be at war. Peking Man had to be protected from the Japanese at all costs.

One might wonder why, in a time of war, when there are certainly many other worries, an invading army would be interested in prehistoric relics of half-million-year-old hominids. But the Japanese were indeed very interested in capturing Peking Man. As the anthropologist Harry Shapiro illustrated in his book *Peking Man*, the Japanese confiscated a skull of Java Man when they took over the island in February 1942. G. H. R. von Koenigswald, who worked there, was arrested by the Japanese. Even though most of the fossils in Java were well hidden, a Japanese search yielded the skull, and it disappeared.

Some years later, after the war, Shapiro asked a U.S. military intelligence officer whether he could help find the missing skull. Soon after the Japanese surrendered, the officer brought him the missing skull. It had been located in the Imperial Household Museum in Tokyo.

In the case of Peking Man, there may have been additional reasons for Japan's desire to possess the remains. The Chinese were proud of their relics and believed that the prehistoric community at Zhoukoudian represented the ancestors of all modern Chinese. These fossils were a source of ethnic and national pride, and symbolized the people's right to their land. It was this nationalistic feeling that the Japanese wanted to quash: By removing the fossils, the Japanese may have reasoned, they could strike at the Chinese heart. Thus the researchers at Peking Union Medical College scrambled to find ways to hide the remains.

Teilhard spent the war years in China in a state of preoccupation and depression. People who saw him noted how sad he looked: He seemed to be contemplating the precarious state of humanity. He was also worried about the future of Father Licent's museum in Tientsin.

Once he finished revising his manuscript, Teilhard brought it to Lucile for typing. She made three copies, and a visiting American diplomat, John Wiley, took one to deliver to Father Edmund Walsh at Georgetown University. Another was sent to the Jesuit headquarters in Rome, with a letter from Teilhard requesting official permission to publish the book. He did not expect authorities to act on his request. Rome now had other worries than Teilhard de Chardin, or so it seemed.

In 1941, with relations between the United States and Japan deteriorating, Americans living in China began to leave. Lucile tried to delay her departure as long as she could. While danger was mounting, she continued to host Teilhard and other expatriates at her house for afternoon tea and to go with him on picnics in the country. But U.S. military authorities were pressing her to leave before the situation became critical: As an American, she could be taken prisoner by the Japanese at any time.

August 8, 1941, was the latest possible moment for Lucile to leave China. The last ship to sail for America had one berth left. Teilhard came to see her off at the train to Shanghai, where she would board the ship, and reassured her that even though he was staying behind, their separation would be only temporary. "God bless you, Lucile," he said in parting, as she wiped her tears.

Teilhard returned sadly to his room. Three days later, he wrote her:

Lucile dear,

May these few lines reach Shanghai before you leave China, to bring you something of the deep of my heart! God bless you again and again for what you gave me since twelve years, and more specially

during these last months!—And may we be together again—very
soon . . .

Be happy—dearest—

Everything is all right,—but I miss you.

Teilhard and Lucile would not see each other again for more than
six years. Their next meeting would be in New York, under very different circumstances.

Teilhard now had other troubles to deal with. His own state of exile
was becoming more severe: His superiors in Rome refused to allow
him to travel to New York to present a paper at the World Congress of
Religions, at which Albert Einstein would be speaking. Teilhard's
attempts to reason with his church had been rebuffed. The Jesuits even
refused to allow him to go to Shanghai to lecture at the Alliance
Française. He was stuck in wartime Peking without permission to
leave for even a short visit outside the city.

He wrote letters to Lucile, who was in Chicago with her family. She
wrote back:

Precious Teilhard . . . Precious Pierre. You are with me so strongly all
the time . . . and you make life always more beautiful. . . . And PT I
am also so grateful to you for being you . . . how much we laughed!!
And the long walks we had, how MUCH I do miss it all . . . but I also
feel sure that we will be together again.

Also in August 1941, Franz Weidenreich departed Peking for New
York, where he was welcomed as visiting researcher in anthropology at
the American Museum of Natural History. The medical college group
wanted him to take the precious Peking Man fossils with him to the
United States, but he could not obtain permission from U.S. and Chinese authorities: It was thought that this would be too dangerous and

that the fossils might be confiscated by some foreign power. Weidenreich tried to convince the U.S. ambassador and the commanding officer of the Marines in Peking to send the fossils to America via official U.S. military channels, but they declined.

After Weidenreich left, carrying with him only casts of the Peking Man fossils, Weng Wenhao, chairman of the China Geological Survey, wrote to the U.S. ambassador, pleading with him to arrange shipment of the fossils to safety. Letters of request were sent by others, but none of these attempts was fruitful. Time was running out.

Before the beginning of hostilities between Japan and China, Kotondo Hasebe, an anthropologist from Tokyo Imperial University, and his assistant, Fuyuji Takai, had visited Peking. According to Pei Wenzhong, their visit was made in preparation for seizing the Peking Man fossils once the Japanese were in control of the city. It has even been suggested that Hasebe and Takai may have been with the Japanese soldiers who took over Dragon Bone Hill back in 1937.

Then the dreaded day arrived. On December 8, 1941—December 7 in Hawaii, the Japanese attack on Pearl Harbor having just taken place, plunging the United States into World War II—the Japanese took over Peking Union Medical College. One of their first acts was a thorough search for the Peking Man fossils. They rounded up several staff members at the college and subjected them to rigorous questioning. But they came up with nothing. Pei was questioned both by Japanese anthropologists and by government agents; according to him, the American administrator of the college was subjected to five days of continuous interrogation. But the team held on to their secret.

Japanese soldiers, accompanied by Professor Hasebe, barged into the department of anatomy at the college and broke open the safe, but found it empty. Angry, they again interrogated Pei, this time very intensively, over several days. They tried to bribe him with the offer of a professorship in Japan, and promised him they would restart the

Zhoukoudian excavation project under Japanese auspices and make him the chief investigator. Finally, they shouted at him that his loyalties were to his race, the Japanese and Chinese family, and demanded he turn against the evil West. When none of this worked, they accused him of being an American agent.

In his memoirs, Pei claimed that he simply did not know where the fossils were hidden, because the Americans, wisely, had not told him. Sometime after his interrogation, Pei was visited by a Japanese agent who identified himself as "George." This man told Pei that his mission was to find the fossils. He harangued Pei for a long time, trying everything in his power to obtain information about the missing relics. The Japanese harshly interrogated captured American personnel, pressuring them to reveal the whereabouts of the fossils. But they, too, were unsuccessful. Hasebe wrote a report on the interrogations and on the importance for Japan of finding the Peking Man bones. His dispatch from Peking was presented to Emperor Hirohito in Tokyo; some scholars believe that because of this report the emperor issued an extraordinary order to his North China Expeditionary Force to make a renewed effort and commit more resources to the search for the remains. In Peking, Chinese and American officials knew that something had to be done about these fossils, and fast.

A FEW DAYS AFTER the Japanese searches began, Chinese workers, supervised by Claire Taschdjian, removed the Peking Man fossils from their hiding place and prepared two large wooden crates for shipping, by military transport, to the American Museum of Natural History in New York. The workers wrapped each of the fossils separately in cotton and secured them with packing tape. Then they carefully arranged the bundles inside the crates. Everything important found at Zhoukoudian—

the remains of forty individuals, as well as stone tools and animal fossils—was now inside two wooden crates. The Chinese were about to entrust their greatest scientific treasure to a foreign power.

Mary Ferguson, an American who remained in China and worked at the medical college, later recounted that she had witnessed the secret packing operation and later saw American military officials moving the two crates across the marble courtyard of the college, through a gate, and onto a waiting vehicle operated by U.S. Marines. They supposedly then transported the crates to their barracks in Peking.

These Marines, who were to be evacuated from China, were charged with the task of bringing the precious cargo to the United States. The two crates were supposed to have been kept at the barracks briefly, while preparations were being made for the Marines, with their equipment, to go by train to the port of Qinhuangdao, on the Bay of Liaodong, within sight of the Great Wall. The *President Harrison* had been due in port on December 8, 1941, and the Marines were to board with the cargo, to be taken across the Pacific to the United States. But the *President Harrison* encountered Japanese warships, ran aground, and never made it to Qinhuangdao.

The crates containing Peking Man never reached their destination in America. They simply disappeared. And for more than sixty-five years, several governments and organizations have tried to locate them or any information about their fate.

Very little is known about what happened. The crates were entrusted to the Marines, but whether they arrived at the barracks is unknown. Various accounts have the Marines taking them out of the gate of the medical college, and others have the soldiers burying them on the grounds of the college or outside it. And there have been unconfirmed reports that the crates were seen at the Marines' compound in Peking. There is also no agreement on what they looked like or on their exact contents. A thick veil of uncertainty obscures everything about the Peking Man relics. It's as if they'd never existed.

Chapter 15

ROME

In November 1942, when another Church official considered the request, Teilhard finally received permission from the Jesuits to travel to Shanghai to deliver his lecture at the Alliance Française. His audience included religious and laypeople of many nationalities—mostly refugees in China, including White Russians and other Europeans fleeing the Nazis—and they did not take well to his speech. He was accused of assuming that evolution was a known fact rather than merely a theory, and some Russians in the audience charged him with embracing communism.

Back in Peking, as the world around him was facing general collapse, Teilhard sought solace in his work. Letters from Lucile arrived only intermittently, and his letters to her fared even worse. From the spring of 1943 until the fall of 1945, Lucile received none of the many letters he wrote her. On November 26, 1941, she wrote him from Chicago: "Pierre dearest,—It is hard to know whether to write or not? If your plans have gone on as they were when I left, you should soon be on your way to America!!" But he was forbidden to leave Peking, regardless of what he planned to do. "I see very few people," Lucile told him, "as I have to take Mother to the Dr. 3 times a week etc. etc. . . . but all this won't last forever . . . and I really don't mind . . . But, oh Pierre, I do miss your talks!! And I DO DO DO miss you!"

On March 12, 1943, the Japanese hauled off to prison camps the remaining British and U.S. personnel left in Peking. The Jesuits were allowed to stay in the city, now almost empty of foreigners. It was a lonely period for Teilhard and for everyone else left there. The following winter was especially cold, with little fuel for heat, and little food as well. The few foreigners remaining were so demoralized that they invited Teilhard to lecture about happiness, which he did just after Christmas, on December 28.

In Rome, while war raged around them, the censors of the Society of Jesus were busy scrutinizing Teilhard's manuscript. On March 23, 1944, one of them wrote a ten-page report, longhand and in Latin, with attached pages of quotations from Teilhard's French, addressed to the Jesuit superior general. The report stressed the reasons that Teilhard's request for permission to publish *Le Phénomène Humain* should be denied. One of the objectionable ideas in the manuscript was Teilhard's treatment of evolution.

The report is kept secret today. The Jesuits apparently feel that it is an embarrassment to the Church, since by now it has accepted some ideas on evolution as valid. But on June 27, 2006, during my visit to the Jesuit Archives in Rome, this document was unintentionally placed in my hands for a few moments before it was taken away. Even today, the Jesuits consider Teilhard de Chardin an extremely sensitive topic. They did not want me, nor do they want anyone else, to see this report, since it harshly criticizes Teilhard for his views. Given that evolution is now taught in schools all over the world as a scientifically supported theory, the fact that they prevented such a towering intellect from publishing because he espoused this theory is still unpleasant for the Jesuits. This document and many others relating to Teilhard's life remain forbidden from view by most people, including scholars within the Society of Jesus.

At the end of August 1945, with the Japanese surrender imminent, Teilhard wrote Lucile:

Dearest,

Just a few lines to let you know that everything is all right here. I am five years older than when you left,—but approximately the same outside, and (I hope) still more the same inside,—especially for you.—In Peking, this long stretch of time was practically uneventful, and rather dull. . . .

Hope that this letter will reach you soon and safely. Evidently, if I have the choice, I will try to go to France via America.

Yours, as before!+++

P.T.

This letter, which arrived in Chicago in the early fall, was the first from Teilhard to reach Lucile in two and a half years. His last letter to her from Peking, written on December 20, was signed, "With much love, P.T.," and she responded in kind. Around the beginning of 1946, she moved to Washington, D.C., and on January 4 wrote him: "Oh Pierre I do hope we meet soon!!! All my love dearest."

At Christmastime 1945, after the Communists had begun their sweep through China, Teilhard finally received approval to return to France. The Jesuits would not leave him stranded in a China that godless Communists were about to take over. Brigadier General William Worton, in command of the U.S. Marines in the country, arranged for Teilhard to fly to Shanghai, and from there he was to take a British ship, the *Strathmore*, which would sail west, not east. While he had been hoping to travel to America to see Lucile, he had no choice, in March 1946, when he was ready to leave. Peking was still a battleground, the Communists were winning, and there were no other options for leaving China.

The Europe that Teilhard returned to was, obviously, not the Europe he had left. Many priests, Jesuits and others, had been arrested by the Nazis because of their ties to the Resistance, and some had been sent to concentration camps. Religion in postwar Europe was not practiced as it had been before the war. The conflict and its atrocities had disillusioned people and weakened their belief in God and in organized religion. This was the beginning of the trend in European society of declining church attendance and Church influence.

All this made the Jesuits even more wary of criticism and more intolerant of views they considered contrary to scripture. When Teilhard left China, he did not imagine that he would need a new place of exile. But soon after he arrived in Paris, it was evident to him that he was not welcome in France, or anywhere on the continent: The Jesuits still considered him too much of a risk. He was an ordained Jesuit priest, so they could not get rid of him; the order to defrock or expel him would have to come from the Vatican, and because of his renown, the process would reflect badly on his order. But the Jesuits would not tolerate his presence where he could shake people's Christian beliefs.

Teilhard was obedient, as was demanded of every Jesuit, and thus Church authorities could order him to leave Europe. Feeling the renewed pressure from Rome in 1946, Teilhard considered moving to Africa. He was interested in the paleoanthropological finds there, and he hoped to visit the famous discovery sites.

In the summer of 1946, he wrote to Franz Weidenreich in New York to inquire about forming a research team including himself and George Barbour to go to South Africa. He might continue anthropological research on human origins that Raymond Dart, the discoverer of the Taung Baby, had begun there in the 1920s.

Weidenreich responded enthusiastically to the idea, and was eager to help his friend. Through his influence, Teilhard was awarded a $2,500 grant from the Viking Fund of the Wenner-Gren Foundation in New York to support his travel and research in South Africa. In the meantime, despite pressure from his order, Teilhard was lecturing in Paris, and his audiences were large and supportive. He spoke of evolution and the noosphere, which attracted a great deal of interest. His view of religion, couched in evolutionary and mystical terms, appealed to a population that had been disillusioned by war. In Paris, all eyes seemed to focus on the courageous priest back from exile in China. He was even nominated for the chair of prehistory at the Collège de France, France's most prestigious research institution. This chair was about to be vacated by his close friend the Abbé Breuil.

On June 18, Teilhard wrote Lucile from Paris:

Dearest,

Pardon me if I did not answer earlier your precious letter of May 21. I am still simply submerged in the Parisian life,—people (every kind of people) calling, telephoning, asking for papers or for lectures all day long. The whole thing is extraordinarily interesting and exciting. But no time is left for writing; and letters, unanswered, are piling up on my table. . . .

Indeed, Teilhard's return to France had been greeted with much publicity: headlines in newspapers and magazines, and the circulation of copies of his unpublished manuscripts. People everywhere wanted to meet him, to interview him, and to hear his ideas. He was a celebrity not only to Parisian intellectuals, but to many ordinary people as well. And to young Jesuits he was a model to emulate. He was living at the editorial headquarters of the journal *Etudes*, on the Jesuit premises on the Rue Monsieur, and whenever he lunched in the garden, young priests would stretch their necks out of the windows above to try to hear his conversations.

Teilhard and Lucile continued writing each other, but both were busy and did not make arrangements to meet. This seems puzzling, when one considers that before the war, both crossed oceans to deliver a lecture or visit a friend. Now, though they had not seen each other for years, neither of them boarded a ship or an airplane to see the other.

On January 5, 1947, Teilhard met outside Toulouse with Jesuits assigned the task of deciding whether he should be allowed to publish his *Phénomène Humain*. After two days of meetings, and scores of objections to various elements of his manuscript, it was apparent to Teilhard that he would be denied permission. In theory, he should have been allowed to publish anything he wanted about science. But the Jesuits contended that the book went beyond pure science, and they used this argument as grounds for rejection. As we now know, that decision had been made in Rome in 1944 when the censor's report was delivered to the superior general. After this new charade, Teilhard returned to Paris in defeat. His days in Europe, he sensed, were numbered. But he felt he had to make another attempt to fight the Jesuit authorities.

Teilhard had one very powerful friend: Monsignor Bruno de Solages, a man close to Pope Pius XII. In September, Solages and a leading Jesuit theologian, Henri de Lubac, met Teilhard at a château near the town of Carmaux, in the French Pyrenees. This château was owned by the monsignor's cousin, who often lent it to him for theological meetings. Over several days, the three men revised the manuscript in an effort to counter the objections raised by the censors in Rome. Ultimately, however, they were unsuccessful in persuading the authorities.

Teilhard continued to lecture widely in Paris. He gave talks at the Sorbonne, at the Musée Guimet (which specialized in ancient Greek and Egyptian artifacts, and Asian art and religion) and the Institut Catholique. He gained wider recognition and was a sought-after speaker throughout Europe. He kept working on *Le Phénomène Humain* in the hope that if he made enough changes, the Jesuits might reconsider. In late 1947, he sent a revised manuscript, with some of the implications about evolution altered, for examination in Rome.

In the meantime, Teilhard received a letter from an acquaintance at Harvard, telling him that Charles Camp, a professor of anthropology at Berkeley, was going to South Africa. Teilhard wrote Camp asking whether he might accompany him. He had waited to arrange the long-anticipated trip because his health was delicate; at sixty-six, he felt weak and he tired easily, and he didn't want to travel alone. Camp agreed to have Teilhard join him, and the two made arrangements for their trip.

During the night of June 1, 1947, while under investigation of his activities by the Jesuits, Teilhard suffered a major heart attack. He was taken to the Clinique Saint Jean de Dieu on the Rue Oudinot, where he remained in critical condition for two weeks, not even able to speak with the many friends who came to visit him. As soon as he was well enough, he was transferred to the clinic of the Augustinian Nuns of the Immaculate Conception, near the forest of Saint-Germain-en-Laye, west of Paris. Here he was slowly nursed back to reasonable health.

While at the clinic, Teilhard had received a letter from his Lyon provincial relaying a request from the superior general in Rome that Teilhard refrain from writing more about philosophy. While Teilhard was not a professional theologian, his writings on philosophy and religion irritated the Church because his arguments could not be readily dismissed. It was intimated that if he did not agree to do so, his works might be placed on the dreaded Index. Recognizing this, Teilhard immediately wrote to the superior general to assure him of his loyalty to the order:

The Father Provincial has recently communicated to me your letter concerning me of 22 August. I have no need to say that, with God's help, you may count on me. I am too surely convinced—and more so day to day—that the world can only be fulfilled in Christ, and that Christ can only be found through an interior submission to the Church.

Teilhard understood that decisive action had to be taken to repair his standing with his order, and he contemplated a trip to Rome to see the general face-to-face. He knew that his dossier at the Holy Office was heavy, and contained what he called "many floating mines." Teilhard had formidable enemies in Rome, but he also had the support of Bruno de Solages. In the spring of 1947, the monsignor had published an article strongly supporting Teilhard and his stand on evolution in the *Bulletin de Littérature Ecclésiastique*. Since everything in the universe evolved, Solages wrote, human thoughts evolved as well; the problem of theology was how to maintain transcendent values in the midst of perpetual flux. More specifically, he commented,

> it is the deep Christian significance of this great scientist's work—of worldwide fame—of this powerful thinker, of this enchanting writer, and, I add, of this *gentleman*, Teilhard de Chardin, to have succeeded in showing, more than any other man, that evolution itself can only be finalistic, that it is advancing toward the spirit, that it can be explained only by the spirit, and that it postulates at the beginning, because it postulates at the end, a transcendent God.

This devoted, powerful supporter was one whom Teilhard sorely needed in his conflict with the Church. When he finally went to Rome, Solages would aid him in his quest. For now, he had to recuperate so he could confront the powerful forces of the Church arrayed against him. He was improving by the day, and exactly six months after his heart attack, on December 1, Teilhard was released from the clinic and resumed his normal life.

During his convalescence, he had written Lucile often. Once he was in the Jesuit house in Paris again, he renewed his desire to visit the United States and see her and other friends. Among these was Rhoda de Terra, who was now divorced from his associate and friend Helmut de Terra; she was living in New York and had been writing Teilhard, and was fast becoming a close friend. Rhoda was a nurse, and Teilhard needed someone to take care of him. In addition, she was a member of the Straus family, partners in the Farrar, Straus publishing house. Her friendship offered Teilhard both the attention of a qualified nurse and the connection to a New York publisher, which might be useful if he ever received permission to publish.

Rhoda made the arrangements for Teilhard's trip to the United States; she also made the necessary contacts in New York so he could do scientific work at the American Museum of Natural History. In mid-February 1948, Teilhard sailed to America. The crossing was very rough, as the Atlantic often is in winter, and still weak, he needed help in moving around in swelling seas. At the end of February, he arrived in New York.

He was on deck as the ship was secured in its berth, and he looked down at the visitors on the dock awaiting the passengers. He was startled, even shocked, to see both Rhoda and Lucile there. He turned to an acquaintance standing next to him and said: "L and R are there; and each one has a car."

Meeting the two women was awkward. Teilhard explained to Lucile that Rhoda was a nurse, and that he needed her help in getting around. Lucile returned to Washington, and Teilhard promised to go see her there. A few weeks later he wrote her: "Dearest . . . It will be good to see the magnolias in bloom, like the cherry-trees of Central Park some time ago. You will see that we will 'rediscover' one another again—as it must be—according to the new times. Finally, I will not arrive in Washington till Wednesday."

Their time together in Washington was pleasant enough, but tensions lingered. Teilhard spent a week there, lodging in the Jesuit house at Georgetown University. He was busy establishing connections with American Jesuits, lecturing, and meeting people. He even went to an embassy party. He did not have time to discuss his feelings with Lucile, and she was deeply jealous of Rhoda. After he returned to New York,

Lucile saw him there. Something, apparently, happened between them then that changed the nature of their relationship. In an undated letter, Lucile wrote him:

> I am very sorry that I "went off the deep end" last evening. I was a poor sport and rotten loser. But I thank you for telling me the truth. I suppose I have really known it for a long time, but in Washington I thought maybe I was wrong and one does not easily accept something that breaks the heart. You say you have not changed toward me: but of course that is not the truth, though you may believe [it]; I assure it is not true.
>
> Years ago when you wrote, not once but several times: "What is born between us is forever . . . I know it." I feared you did not know (I think no one can *know*) but I so much wanted to believe and because of all the circumstances, and also even our ages, made it seem very possible, and so I built my life upon it.

More conciliatory letters followed, in which they both expressed love for each other, and late in the spring, Teilhard returned to Paris. They had not had the heart-to-heart conversation they needed. They knew that they had to see each other again, whether it meant putting things back together or permanently breaking up. The same considerations weighed on Teilhard's relationship with the Church.

WHILE HE HAD NEVER VISITED THE CITY, Teilhard had always considered Rome the center of Christian civilization—he called it "the axis of humanity." And it was Rome that would be the setting for his final showdown with the Church—and, though he did not know at the time, also with Lucile.

In early October 1948, Teilhard traveled to Rome to face his accusers—just as Galileo had done to face the Inquisition some three centuries earlier. The Jesuits had agreed to meet with him in person to discuss his lectures and writings that were so abhorrent to them. Teilhard wrote to Jeanne Mortier, a friend he had met in Paris in 1939 when she had helped with one of his presentations at the Musée de l'Homme:

Rome, October 8, 1948

Dear Friend,

This to thank you again for having seen me off so kindly.—and to assure you that my journey has been perfect. I finally opened my eyes only at Brig—at the entrance to the Simplon. I was thus able to enjoy the marvelous descent into Lake Maggiore. But darkness befell us before we entered Umbria.—which is a shame. A friend waited for me on the platform at Rome at midnight.—so that I should get to my bed without any difficulty. This was my first morning, on the 4th. Since then I've begun my Roman education. Altogether, a welcome marked by a bit of curiosity—but overall sympathetic—even more so given that *here* I have no *real* friends. I found my lodgings not at the [Pontifical] Biblical Institute (no room), but in the general home itself (within a community apart, that of "writers"). That is to say that I am at the immediate area bordering the Vatican—a few hundred meters from Saint Peter's, whose creamy dome dominates the gardens against which our home stands. The community here is very heterogeneous (both of nationalities and of occupations) and too large to be truly pleasant.

Shortly before writing to Mortier, Teilhard had written a few lines to Lucile Swan. A month earlier, while he was at his brother's estate near Les Moulins in the Auvergne, Lucile had been in Paris. They did not meet. He was unsure of her feelings toward him, yet still felt a strong attachment to her. When he heard that she was in Paris, he wrote her:

Les Moulins, September 3, 1948

Dearest,

Such a thrill to receive your letter from Paris!—And such a happy feeling that you like it!—Yes, let us hope that this new contact with so many old things will rejuvenate you, and show you *your* line of life, and, *ipso facto, your* God. . . .

A bientôt, in any case.

God bless you, dearest!

Lucile soon left Paris and headed for Switzerland—not the Auvergne—and Teilhard went to Paris and from there left for Rome. Before leaving, he heard from Lucile in Bern. She might come to see him in Rome, she said. So now he wrote her again:

Roma, October 7, 1948

Dear L.,

These few lines to tell you that I have arrived easily and safely here, last Sunday, at midnight!—I am located a few hundred meters from la place St. Pierre [in French in original],—at the very fringe of the Vatican!—Got a very charming welcome. But it's too early for having any definite prospects concerning my affairs.—Your letter from Berne reached me in Paris just before my departure. Thank you for everything!—When you are here, let me know. The simplest thing, to start, should be that you call here some day when going to visit St. Peter (better in the morning,—f.i. after 10 a.m. and before noon) and ask for me: there are little reception rooms downstairs (like in Rue Monsieur),—and, in addition, good elevators. . . .

Have a good time in Berne!

Yours as ever

Teilhard

They had difficulties meeting. She had been hurt, and her feelings toward him had cooled, which made her efforts to see him more casual. But Teilhard wanted very much to see her. Later in October, he wrote her a note:

My dear L.,

This afternoon I must be back here, at 4 p.m., to meet an influential colleague.—And tomorrow I *may* have to stay here, more or less the whole day, waiting for possible developments.

The best is perhaps that you should pass here tomorrow morning, between 10.30 and 12. And then I might perhaps go to the Flora at 4 p.m., if I am free.

Would you kindly make sure at the desk of your hotel that the Paris train (via Simplon) is still leaving at 7 *a.m.*? You could tell me, tomorrow morning.

Your

P.T.

If you don't come tomorrow morning, I will wait for you here (as much as I can) in the afternoon, after 3 p.m.

They did meet briefly. But soon Lucile left for Ethiopia. As an artist, interested in the bright colors and the quality of light in that country, she had always wanted to visit. Teilhard remained in Rome awaiting the Jesuits' answers to his requests. His relationship with Lucile had lost all its warmth and excitement; it was virtually at an end.

After some time, Teilhard was informed that the superior general was too busy to see him, and that in any case, the Jesuits were still waiting for the arrival of two copies of *Le Phénomène Humain*, which had been sent out for review by clerics. Teilhard remained in the city, waiting for his audience with the superior general. He often sat at a café in the Piazza del Popolo, staring at the high obelisk brought from Egypt and wondering why it was necessary to place a cross on top of it. Couldn't the ancient obelisk be left alone as the pagan symbol it was?

One evening, Teilhard was invited to a cocktail party in a palazzo adjoining the church of San Luigi dei Francesi, the French church in Rome. There, he stared across a crowded room to meet the eyes of his archenemy in the Holy Office, the Dominican theologian Father Réginald Garrigou-Lagrange. It was he, Teilhard knew, who was behind many of the problems the Vatican was creating for the Jesuits because of Teilhard's ideas. Teilhard turned to an acquaintance and said, "This is the man who would like to see me burned at the stake." Moments later, Garrigou-Lagrange pushed his way over to Teilhard, shook his hand, and began to speak to him about the Auvergne.

Pope Pius XII was not in Vatican City then—he was still on vacation at the papal retreat of Castel Gandolfo, in the Alban Hills about twelve miles southeast of Rome. A friend of Teilhard's had a private audience with the pope around this time and mentioned to him Teilhard's positive Christian influence in Peking, which was furthering the interests of the Church in the Far East. The pope replied, "I know that Père Teilhard is a great scientist, but he is not a theologian. In one of his essays he speaks of 'resolving the problem of God.' But for us there is no problem." Apparently, the Vatican had no understanding of Teilhard de Chardin and his ideas. The Jesuits could have offered him some help, but they did not.

Even though Teilhard was told that the manuscript copies had not come back from review yet, the superior general agreed to see him. The superior general, Jean-Baptiste Janssens, showed genuine interest in Teilhard and his predicament, but was not encouraging about his requests. Janssens told Teilhard that he would not be allowed to take the chair he had been offered at the Collège de France, because accepting such a position would create a scandal. The Jesuits worried that this highly prestigious position would give Teilhard even greater visibility than he already enjoyed. And since his views on evolution and original sin were taboo, his elevation to the post and the pulpit it would provide for these views would create controversy within Jesuit circles—and would have serious repercussions in the Holy Office as well.

Janssens was of no help in securing publication permission from the Jesuits or the Vatican. The process for obtaining such permission was complicated, and in the hands of semi-independent censors whose reports were delivered to high officials at the Vatican and to the Jesuit superior general. Trying to interfere with the process would not be in the interest of the Society of Jesus, as the Vatican would not look kindly on such meddling.

Teilhard stayed in Rome, and finally the answer to his request to publish *Le Phénomène Humain* came from the censors: No. By early November he understood that none of his requests would be granted. Dejected, he boarded a train for Paris.

His trip to Rome had failed. His deepening conflict with the Church and with his own order had not been resolved. He was not allowed to accept the important academic position that he had been offered, which he greatly desired, and his manuscript—on which he had labored so intensively and for so long—had been rejected. His relationship with Lucile was irreparably damaged, too. Although they corresponded until his death, their friendship was crippled. This trip to Rome marked a somber turning point in his life. From then on, nothing held the same promise as before.

Teilhard resumed his life in France, and spent the following year and a half lecturing on science, speaking to religious audiences, and visiting his relatives in the Auvergne. At the invitation of his friend Jean Piveteau, who held the chair of paleontology at the Sorbonne, he gave a series of lectures at the university. In February 1949, he spoke there on "Man's Sense of and Place in Nature." In August, before going to his brother's in the Auvergne, he gave a series of lectures at the Jesuit center on "Man and the Definition of Species."

In March 1950, Teilhard delivered lectures on "The Phases of a

Living Planet" and reviewed books on evolution and the discovery of hominid fossils. That same month he contributed a series of articles to the journal *Les Nouvelles Littéraires*, in which he discussed questions of human evolution and the transmutation of inanimate matter into early life forms. While he was permitted to write papers about science, these articles enraged the Vatican.

On May 22, Teilhard was elected to membership in the French Academy of Sciences, in the paleontology section. He hoped that this new honor would offer him some protection from the attacks against him from Rome and the Vatican. But the honor made the Church even angrier with him, and there were further efforts to isolate and intimidate him, and to punish anyone who supported him and his views. That year, a book designed to ridicule him (its title translating to "The Redemptive Evolution of Père Teilhard de Chardin") was published anonymously in France. In the wake of Pope Pius XII's encyclical *Humani generis*, issued on August 12 and dealing in part with evolution, Jesuit academics who had espoused Teilhard's ideas, among them his friend Henri de Lubac, were ordered by the Vatican to leave their positions.

Chapter 16

AFTERMATH

Feeling renewed pressure and the potential for another exile, Teilhard was ready to visit South Africa. He spent several months studying anthropological discoveries made there, and since his heart attack had cost him the opportunity to travel with Charles Camp, he asked George Barbour to join him on the trip. After his heart attack, it was even more important that Teilhard travel with someone who could help take care of him. Barbour agreed to meet him in South Africa.

Since Teilhard might not return to France once he left this time, his friends persuaded him to bequeath his papers to Jeanne Mortier, so that they might be published one day—something that would not be possible if they were in the possession of the Jesuits. Teilhard agreed and made the appropriate arrangements. On July 12, 1951, accompanied by Rhoda de Terra, he left from England aboard the *Carnarvon Castle*. They arrived at Cape Town in late July, and took the train to the Transvaal. Barbour was there waiting for them. He gave Teilhard reports on South African excavations, which he read carefully. Though he was weak, he felt better being in the field again, and he was even able to do some moderate climbing to inspect paleontological sites.

Teilhard felt well enough to visit the locations where discoveries had been made in the years since Peking Man was found, such as Sterkfontein, site of an important *Australopithecus* discovery by Robert Broom in 1936. On August 7, Teilhard traveled by jeep with his host, Dr. Clarence van Riet Lowe, to a ridge above a cave where work was being done to recover hominid fossils. He lowered himself by rope into the cave, as he had done so many years earlier at Cro-Magnon sites in Europe.

Teilhard stayed on in South Africa after Barbour left in September. He had news from France, sent him by Jeanne Mortier, that the Jesuits had a new provincial in Lyon. Teilhard, ever hopeful that his situation might improve and that he would be allowed to return to France, immediately sent the new superior a letter to see how he felt about various matters. But his optimism was once again dashed. The provincial, Father André Ravier, politely informed Teilhard that, given the way Rome viewed him, if he were to reenter France, he would have to be confined to a retreat house where he would live under strict surveillance. The choice for Teilhard was unequivocal and dreadful: new exile or confinement. The provincial advised strongly that Teilhard find himself a scientific position in the United States.

Ravier also told Teilhard that he could write whatever he wanted about philosophy, and promised that he would read it, but reiterated that publication was of course a different story, requiring permission from Rome. Teilhard, he suggested, might write a letter to his general, reaffirming his loyalty to the order. Teilhard wrote the letter, promising that he would never again propagate his philosophical ideas without permission. But he was still an exile, more than he had been before, and he was not allowed to return to France.

On October 18, 1951, Teilhard left South Africa for the United States. He arrived in New York on November 26, and rushed to visit the director of anthropological research at the Wenner-Gren Foundation, Paul Fejos, who had said he would try to obtain a position for him with the Foundation. Teilhard's problem of finding employment in the United States was solved almost immediately, much to his relief, when Fejos offered him a research position. Teilhard lodged at the Jesuit house adjoining the Church of Saint Ignatius Loyola at Park Avenue and Eighty-fourth Street. The members of the order living there welcomed him with great respect and interest. And while he may not have been happy to be in exile from France, he found contentment in his new job.

At the time, Lucile was living in the city, but Teilhard saw her only rarely. His letters to her during this period are not as loving as before; he addressed her simply as "Dear Lucile," as he had done when they first began their correspondence. Lucile was upset by Rhoda's closeness to Teilhard; again she felt hurt, and she carried her resentment about years spent in an unsatisfying relationship.

Their difficult friendship weighed heavily on Teilhard now that he and Lucile were again in the same city. Years later, when she was in her nineties, Rhoda was asked about Lucile Swan. She smiled and said, "Lucile just didn't know how to play the game. [She] thought he wanted a girlfriend; he wanted a mother, so I mothered him." It is unlikely that Teilhard gave Rhoda the intimacy he could never bring himself to give Lucile; Rhoda had been married to a man who was still his close friend—something that would have made an intimate relationship between them even more complicated and contrary to Teilhard's moral code.

To cheer Teilhard up, new travel was on the horizon. The Wenner-Gren Foundation was interested in renewed paleontological research in Africa, and it arranged to send Teilhard to South Africa. He was working hard preparing for the trip, part of the Foundation's Operation Africa, and gave talks about the project in Philadelphia and New York. He was revising his ideas about evolution, and now concentrating on the concept of speciation, the formation of new species. Teilhard hoped to bridge anthropology and biology by defining "hominization"—the development of hominids—in terms of the biological process of speciation. He identified socialization in early societies as a form of speciation in humans. In this view, humans differed from other animals in that socialization was a fundamental element of their becoming a species.

Teilhard took the bold move of requesting permission from his provincial in Lyon to travel through France while on his way to South Africa. He explained that he would be traveling through London, and thus requested permission to cross into France for a few weeks' visit. As he expected, the answer was no. Teilhard was forbidden even a short visit to his home country. In his response, Father Ravier pointed out that Teilhard had not stayed away from France long enough, and that thus there was no reason for him to return now.

In early July 1953, Teilhard embarked on his trip to South Africa. He stayed for a short time in Cape Town, examining new hominid fossils that had been found over the past few years. He then went to Johannesburg to see Raymond Dart's newest discoveries, which included a lower jawbone of *Australopithecus*. He had been planning to visit Louis Leakey's excavation sites in East Africa, but because of the Mau Mau uprising against the British colonial powers in Kenya, these sites were closed. Instead he visited the British archaeologist J. Desmond Clark's Livingstone Memorial Museum and a site near Lusaka, both in Northern Rhodesia, and cruised down the Zambezi River. From Johannesburg in mid-August, he wrote to Jeanne Mortier:

In Cape Town, we found magnificent blue skies (snow on the mountaintops)—which have not left us since. And now, despite the altitude here (around 2000 meters) it's springlike weather.—At moments, I feel as if I've returned to northern China.—Jo'burg resembles an American city. . . . I love it here. Scientifically speaking, everything is developing well. I've seen for the first time the sites of the *Australopithecus*, close to here; and I spent last week 250 kilometers north of here, in another series of famous sites—where I learned a lot. The geologists and prehistorians here are charming; and I've begun to understand clearly the local problems.——and the ones relevant to the W. Gr. Foundation.——Finally, the altitude doesn't make me tired; and, nervously, I feel better than I have felt in three years. I like seeing in this an indication that the Lord watches over me and still expects something of me. . . . In the scientific "excitement" [in English in original] (which moves me, albeit in a less vivacious way than it once did), it seems to me that my essential vision of the world is finding a means of clarifying itself and intensifying ever more. I am meditating now on an essay on *The Triple Reflection* (reflection on the life of an individual man, reflection of humanity on itself, reflection of God on the thinking humanity), which I believe will go further than the preceding one [the 1948 essay "Three Things I See, or A Weltanschauung in Three Points"].

In Paris, Mortier had had the idea of broadcasting Teilhard's forbidden writings on the radio, and she wrote him of this. On September 10, Teilhard answered:

> Concerning your project of broadcasting certain texts of mine—if this project is still viable—I think it would be better to refrain. . . . At this moment, according to a letter I've received, Rome remains defiant. . . . It is better to concentrate on fundamental work—without any provocation.

Teilhard returned to New York, and from there wrote once more to Ravier, who in his earlier rejection had hinted that Teilhard might apply for permission to visit France the following year. This time when he applied, Teilhard received the provincial's approval. He was delighted to be able to return home, and his uplifted mood could not be soured even by the news, in November, that the Piltdown Man "discovery" had been proven a hoax. He was surprised that Charles Dawson had fooled so many people, and was embarrassed at having fallen prey to his deception. Despite later accusations by Stephen Jay Gould that he was somehow involved in this fraud, Teilhard was without doubt innocent in this matter.

By this time, Teilhard had elevated the concept of evolution to the pinnacle of both science and religion. To him, the integration of the concept of time within the idea of space, as in Einstein's relativity, made evolution the prime mover of the universe. In his 1953 essay "Le Dieu de l'Evolution" ("The God of Evolution"), he observed: "A century ago, evolution, as it is called, could still be regarded as one simple and local hypothesis which had been formulated and which was used in connection with the problem of the origin of species, and particularly the problem of human origin. But since then it must be recognized that it has expanded in every direction and now speaks to the totality of our experience." Teilhard's God is the creator of our immense universe, and he makes the universe move forward through the process of evolution.

In June 1954, Teilhard sailed to France. He went to Sarcenat, but because only one member of his family, his brother Joseph, was still alive, Teilhard did not stay long. He proceeded to the Périgord, where he met the Abbé Breuil. They visited the cave of Lascaux with its impressive prehistoric art, which Breuil had authenticated by analyzing the color material, as he had done before in other caves.

Teilhard spent the rest of the summer lecturing in Paris, and again stirred up controversy. The Jesuits who attended his talks were hostile to his ideas, perhaps because in his absence from France the Church had been expending its energies trying to discredit him. Trouble was brewing for Teilhard despite the brevity of the visit: The religious in Europe simply could not tolerate Father Teilhard de Chardin. On July 31, he received a letter from Jesuit headquarters in Rome ordering him to return at once to the United States.

Despite his obedience, which other Jesuits have described as "exemplary," Teilhard's mere presence in Europe was disturbing to the Jesuit leaders. Now old and weak, he nevertheless had the strength to obey his church and travel. The Jesuits had neither pity for him nor understanding of his wish to spend his remaining days in his beloved homeland, and they sent him to his final place of banishment.

Realizing that he would never be allowed to fulfill what he saw as his scientific and philosophical mission, Teilhard returned to New York. He was sentenced to living out his days in a foreign land: His ideas proved too much for his faith. Still, he continued to formulate them more precisely, and wrote a treatise about his hero Galileo. But the Jesuits prevented him from publishing the books he had written.

On March 30, 1955, Teilhard wrote his last letter to Lucile Swan:

> Lucile dear,
>
> Merci, tant, for your letter (March 28)
>
> Yes, stupidly enough, I am still nervous,—more nervous than I would,—than I should be.
>
> And, at the same time, I need definitely your presence, your influence, in my life.
>
> I hope (I am sure) that things will gradually settle, "emotionally" speaking.—In the meantime, and as a minimum (or as a provisional "optimum") we might try to see each other at the rate of two–three times a winter.—In any case, we know, both of us, that we "are always here" for each other.—Phone me any time you like.—I will let you know anything important or interesting which may happen to me. And I shall certainly see you before I leave New York for the summer.—My plans are still vague, on account of this awful question of "permanent visa" which I have not got so far!
>
> God bless you for all you gave and give me!
>
> Yours, very affectueusement,
>
> Pierre

Days later, on Easter Sunday, April 10, at the age of seventy-four, Pierre Teilhard de Chardin died in New York. He had attended High

Mass at Saint Patrick's Cathedral that morning, and in the afternoon took a walk in Central Park. He spent the rest of the day with Rhoda de Terra and her daughter. They had planned to attend a concert that evening, but then went to Rhoda's apartment instead. At about six p.m., Teilhard was standing with his friends when he suddenly fell to the ground. He regained consciousness and asked where he was and what had happened. Rhoda immediately called a doctor, but by the time he arrived, Teilhard was dead. He had suffered a massive cerebral hemorrhage.

Teilhard's body was embalmed and brought in a coffin to the Church of St. Ignatius Loyola. On the following Tuesday, a requiem was chanted for him. Among the few people attending was Lucile Swan. The coffin was taken to the Jesuit novitiate of St. Andrew, ninety miles north of New York City, along the Hudson River. The ground was still partially frozen, so the coffin was kept in a vault for several days, until a grave could be dug. When Teilhard de Chardin was finally buried, none of his friends or associates was present. He had died too far from the cultural center in which his ideas had gained an audience in his lifetime; but within a few short years, this was to change.

Soon after his death, Jeanne Mortier arranged for Teilhard's books to be published. Because he had bequeathed the manuscripts to her, the Jesuits could no longer prevent their publication. Thus, *The Phenomenon of Man* (1955), *Letters from a Traveller* (1956), *The Divine Milieu* (1957), and *The Future of Man* (1959) all appeared in France, and soon after elsewhere, to great acclaim. In death his ideas attracted interest, and today there are Teilhard de Chardin societies in several countries. People continue to read and appreciate his theories and thoughts, and his view of evolution has been received with interest. In 1996, Pope John Paul II celebrated the fiftieth anniversary of his priesthood with the book *Gift and Mystery*, in which he quoted from Teilhard's "Mass on the World," written in the Ordos Desert in 1923.

In his book *Mankind Evolving*, the prominent geneticist Theodosius

Dobzhansky summarized Teilhard's influence on modern science and thought:

> To modern man, so forlorn and spiritually embattled in this vast and ostensibly meaningless universe, Teilhard de Chardin's evolutionary idea comes as a ray of hope. It fits the requirements of our times. For [as Teilhard wrote in *The Phenomenon of Man*] "Man is not the centre of the universe as was naively believed in the past, but something much more beautiful—Man the ascending arrow of the great biological synthesis. Man is the last-born, the keenest, the most complex, the most subtle of the successive layers of life. This is nothing less than a fundamental vision. And I shall leave it at that."

Chapter 17

THE FOSSIL RECORD CONTINUES

In the half-century since Teilhard de Chardin's death, research on our human origins has expanded tremendously. And while the world has never again seen an excavation project on the scale of Zhoukoudian in its heyday, today's researchers use an integrated approach, incorporating methods from many fields, which makes them more effective at finding and analyzing fossils. Still, it is a daunting task to find fossils in the ground. One needs to know where to look and how to search. The last few decades have brought us a new crop of highly motivated fossil hunters, and their finds keep growing and filling the gaps in the chart of our ancestry.

Most new discoveries have been made in Africa—which now appears to be our species' ancestral home. Our own genus, *Homo*, appeared on the plains of that continent between 2.5 and 3 million years ago. The new hominids, *Homo habilis*, evolving from the more primitive australopithecines, inhabited the African savanna concurrently with the australopithecines for a period of several hundred thousand years, until these earlier creatures died out. The robust australopithecines, of the kind discovered by Robert Broom in Kromdraai, South Africa, in 1938, lived for perhaps a million years, coexisting in Africa with both *Homo habilis* and the later *Homo erectus*, before disappearing. The gracile australopithecines, first discovered by the Leakeys, are believed to have been our ancestors through *Homo habilis*. Eventually, they too became extinct.

Olduvai Gorge, in the eastern Serengeti Plain and the famous wildlife park of Tanzania, was the site of the great discoveries by the Leakeys and their coworkers. Louis Leakey first found stone tools here in 1931, but the search for hominids resumed in earnest only in 1959, after Mary Leakey found a skull while walking her dog. The skull was that of a creature with a small braincase and large teeth—which therefore did not qualify to be called *Homo*. Louis Leakey named it *Zinjanthropus boisei* (East African Man). It was an australopithecine now dated to about 1.75 million years ago. This skull was much studied, and it made Olduvai Gorge famous. Its species has since been renamed *Australopithecus boisei*.

Not far from that location, Louis's son Jonathan found fragments of a larger skull, with a cranial capacity of about 650 cc, as compared with the australopithecines' average 500 cc and modern humans' 1,400 cc. He concluded that this hominid was a member of our own genus—a decision later questioned by some experts, since the dividing line between *Australopithecus* and *Homo* has never been clear—and named it *Homo habilis*. This was the putative first member of our own genus. These hominids walked upright, more often and better than did the australopithecines, and they had larger brains and were more humanlike.

In 1974, Donald Johanson was searching for fossils in Hadar, a region in the Great Rift Valley in Ethiopia, when he discovered an almost complete hominid skeleton—an unparalleled discovery for its age: 3.2 million years. Within the two-year span of the project he was involved in, from 1973 to 1975, a total of 250 hominid fossils were found here, leading to the naming of a new species, *Australopithecus afarensis* (for the Afar people who inhabit this region of Africa). These fossils, and others discovered farther south, range in age from about 3 million to 3.8 million years. The hominids these fossils represent were shown to have changed very little anatomically over their almost million-year existence, thus proving that they were extremely well adapted to their environment.

The almost complete skeleton that Johanson found was determined to have been that of a woman. The tape deck in the tent at the excavation site was blaring the Beatles song "Lucy in the Sky with Diamonds" at the time of the discovery, and Johanson and his colleagues decided to name their skeleton accordingly. "Lucy" has attracted worldwide attention and is undoubtedly the most famous fossil ever found.

Johanson has described her: "From the neck up, chimpanzee; from the waist down, human." Moreover, "for all their apeness, Lucy and her kind did share the first human evolutionary marker: They walked. They kept on walking, beautifully adapted to their African environment, for another million years."

In 1978, Mary Leakey and her team made a stunning discovery. In a field of volcanic ash, dated to 3.5 million years ago, in a place named Laetoli, not too far from Olduvai Gorge in Tanzania, she discovered footprints. These were covered with ash shortly after the hominids that made them had departed. This find provided crucial evidence that *Australopithecus afarensis* walked upright. The Laetoli footprints constitute unique and direct evidence for the bipedalism of hominids that lived such a long time ago. Until this discovery was made, there had been only indirect evidence for bipedalism, based on the structure of the skeleton as inferred from bone fossils.

In 1984, near Lake Turkana in the Great Rift Valley in Kenya, Alan Walker, Richard Leakey, and their team discovered the "Nariokotome Boy," a fossil representative of the African *Homo erectus*. Thus the species to which Peking Man belonged was discovered in Africa. Some anthropologists thought that the find should be identified as a separate species, *Homo ergaster*, the name given to a species of which a single mandible was discovered in 1975. But Nariokotome Boy's discoverers believed that this juvenile hominid was of the same species as

Java Man and Peking Man—that is, a member of the long-lived species *Homo erectus*.

On April 13, 2006, the journal *Nature* reported on one of the latest discoveries in human evolution. An international research team headed by Tim White of the University of California at Berkeley had uncovered fossils in the Middle Awash Valley of Ethiopia that were dated to about 4.1 million years ago. The fossils were said to represent a primate intermediate in its anatomy between the earlier *Ardipithecus ramidus* (First, or Root, Ground Ape) discovered by White in the Middle Awash in 1992 and dated to around 4.4 million years ago, and the later *Australopithecus afarensis*, the species to which Lucy belonged. The newly discovered relics are the earliest known for the species *Australopithecus anamensis*, first discovered by Mary Leakey and dated to 4.2 to 3.9 million years ago.

These findings seemed to support the hypothesis that *Australopithecus anamensis* was a direct ancestor of *Australopithecus afarensis*. While the earlier *Ardipithecus* was more like an ape and had a smaller brain, it too walked upright.

Also published in *Nature* in 2006 was news from Dikika in Ethiopia, where a research team headed by Zeresenay Alemseged had discovered skull and bone fossils of a girl about three years old, dated to 3.3 million years ago and a member of the same species as Lucy, *Australopithecus afarensis*. The remains of this child, dubbed "Lucy's Baby" although she lived 100,000 years earlier than her "mother," taught us new things about evolution. Because much of her skeleton was preserved, scientists have been able to confirm that this species indeed walked upright. But Lucy's Baby also showed distinct signs that the species' upper limbs were used for climbing trees. Thus, as far as locomotion, this species was somewhere between humans and apes. And again in 2006, G. Philip Rightmire and his colleagues reported on discoveries of *Homo erectus* in Dmanisi, in the Caucasus region of the Republic of Georgia, that had been made over previous years. The fossils resembled those of the African *Homo habilis* but were determined to belong to *Homo erectus*, thus extending the known range of the later species to the outskirts of Europe.

The previous year, research on hominids that lived much further back in time was published, addressing a hominid discovery in Chad in 2001. *Sahelanthropus tchadensis* was the first African fossil human ancestor to be found in the north central part of the continent. The Chad fossils have been dated to about 7 million years ago. This was an upright biped, and its great antiquity demonstrates bipedalism existing even at this distant time in the past. Scientists were led to conclude that this hominid likely evolved soon after the divergence of the human lineage from that of chimpanzees.

Another hominid close to the last common ancestor of humans and chimpanzees is the 6 million-year-old *Orrorin tugenensis*, discovered by a French team led by Brigitte Senut and Martin Pickford in 2000 and appropriately nicknamed "Millennium Man." This creature, its discoverers believe, was bipedal. In fact, they claim that an analysis of its femur implied that it was even better adapted to walking upright than was *Australopithecus*.

WITH ALL THESE NEW DISCOVERIES, what was Peking Man's role in the story of human evolution? Franz Weidenreich was a strong proponent of the theory of multiregionalism. According to this view, Peking Man locally gave rise to *Homo sapiens* in Asia by continuous evolution. The Chinese, according to this hypothesis, are the descendants of Peking Man. We know from the fossil record from Zhoukoudian that *Homo erectus* lived there for at least 260,000 years (from 670,000 to 410,000 years ago), and perhaps this finding (or a similar one, since these more exact dates were derived later) inclined Weidenreich toward this view of evolution, at least in China.

According to the multiregionalism theory, Java Man gave rise to Australian aborigines; Heidelberg Man (*Homo heidelbergensis*), which evolved from *Homo erectus*, became modern Europeans; and the African *Homo erectus* (or *Homo ergaster*) evolved into modern Africans. The multiregionalism theory has been likened to a candelabrum. Imagine a candelabrum that has its base in Africa and branches emerging there and extending into Europe and Asia. *Homo erectus* left Africa around a million years ago and followed these branches into the regions of the Old World. At each "candle," each end of a branch, *Homo erectus* evolved (sometimes through intermediate species) separately into a different race of *Homo sapiens*.

All serious theories about human evolution posit a beginning in Africa, so the question really is: When did we leave Africa? According to multiregionalism, we left the continent early and then evolved into *Homo sapiens* concurrently at various locations in Europe and Asia. But modern views of human evolution tend to favor a different theory, called "Out of Africa." It posits that *Homo erectus* was our ancestor. The

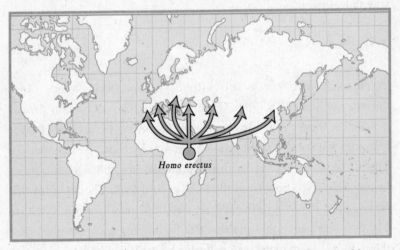

The multiregionalism theory. *Homo erectus* migrates from Africa into Europe and Asia, and evolves in various localities into *Homo sapiens*.

Homo erectus that lived in Africa (*Homo ergaster*) evolved in Africa into *Homo sapiens*, and then this anatomically modern human migrated into the Middle East and from there to Europe and Asia.

According to Out of Africa, then, modern humans left Africa late, and then replaced the older *Homo erectus* populations (and their descendants) already existing in the Old World. In this sense, the theory might well be called "Out of Africa Again."

In 1996, Carl Swisher and others published an article in the journal *Science* in which they dated some *Homo erectus* fossils found in Java to as recently as 30,000 to 50,000 years ago. If these dates are correct, then the period in question saw no less than *three* human species inhabiting Earth at the same time: *Homo erectus* (the species of Peking Man, but much younger than the finds at Zhoukoudian) in Asia, Neanderthals in Europe and the Middle East, and modern humans in Africa, Europe, and the Middle East. And if *Homo erectus* was stable enough as a species to last so long—close to 2 million years—then why did it disappear 30,000 years ago? Were the last remaining Asian *Homo erectus* populations

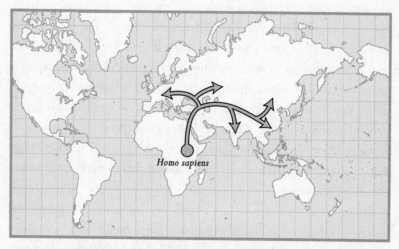

The "Out of Africa" theory. *Homo sapiens* evolves in Africa, migrates into Europe and Asia, and replaces other hominids already established there.

replaced by modern humans in the same way that Neanderthals in Europe were replaced by anatomically modern humans?

PERHAPS THE GREATEST QUESTION in anthropology is: When did we become human? When did we evolve into feeling, conscious beings with the capacity for symbolic thinking and everything that comes with it—language, art, literature, mathematics, science, philosophy, culture, and an economy?

Our "aha" moment of consciousness, when human civilization began, seems to have taken place about 30,000 years ago in Europe. So while our ancestry is in Africa, our symbolic thinking began in a European cave when, a thousand generations ago, a Cro-Magnon painted the walls of a cave with representations of animals, and symbols whose meaning remains a mystery. (It should be noted that rock art and other decorations, some found outside Europe, could also be interpreted as the result of symbolic thinking.)

A bank vault in Germany holds a collection of beautiful carved statuettes: horses and mammoths and humanlike figures. Dated to 32,000 years ago, they were discovered in the early 1930s in the Vogelherd cave in southwestern Germany. One item is particularly interesting: a horse carved out of mammoth ivory, two and a half inches long, on which can be seen markings and evidence that it was handled over a very long period of time. Scientists believe that this and similar objects represent symbolic thinking, and that the necessary communication among members of the group that handled such objects must imply the use of language.

Richard Leakey believes that a rudimentary form of language was spoken by *Homo habilis* in Africa. He even thinks that australopithecines may have had some form of verbal communication. According to Leakey, *Homo erectus* produced language, increased vocabulary,

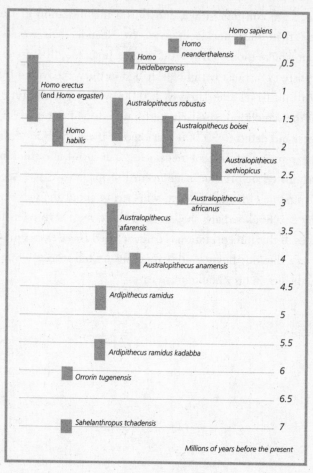

Timeline of various extinct hominids whose fossils have been discovered, and modern humans.

and perhaps developed sentence structure; then, from 30,000 years ago onward, emerged the roots of modern languages. Anthropologists including Ian Tattersall of the American Museum of Natural History believe that neither the Neanderthals nor earlier hominids possessed language. Language and other symbolic thought, these researchers say, came about only with *Homo sapiens* and coincide with the appearance of European cave art about 30,000 years ago.

If they are considered art, and thus a manifestation of symbolic thinking, then shell necklaces made by Neanderthals would imply some humanlike thought. An article in the June 23, 2006, issue of *Science* reported on shells with holes drilled in them recovered from the Skhul cave in Israel and dated to 100,000 years ago, and on a single shell of the same kind, similarly perforated, found at Oued Djebbana in Algeria and dated to 90,000 years ago. If these shells indeed represent the remains of ancient necklaces, then symbolic thinking may have originated before the magnificent European cave art. According to Jia Lanpo and Huang Weiwen, who wrote about their experiences working at Zhoukoudian, shells found there may have been used in necklaces. If this interpretation is correct, then *Homo erectus* might also qualify as a thinking being. And perhaps he did tell stories while sitting around the fire in his Zhoukoudian cave.

Chapter 18

WHAT REALLY HAPPENED TO PEKING MAN?

Despite the many other discoveries that have been made in paleo-anthropology since World War II, the world will not forget Peking Man. The recently freer communication between China and the West is helping in the renewed search for the long-missing fossils.

Unfortunately, however, the sixtieth anniversary of the end of World War II saw a recrudescence of national anger and mutual distrust between China and Japan, and a deteriorating diplomatic relationship between the two nations. This makes more difficult the exchange of information between them that is essential for an effective search for the Peking Man fossils. The official Chinese investigation into the fate of the relics, announced in September 2005, has therefore been confined to information available in China.

Of the many theories proposed over the decades about the fate of the fossils, some published and others propagated by word of mouth, we can certainly discount one. This is the belief that the crates sank at sea when the ship carrying them to the United States was attacked by the Japanese. We know that this theory is wrong, unless the containers were loaded onto another ship that then sank; records show that the *President Harrison* never made it to port in the first place. The ship, en route to China from Manila, was being pursued by Japanese warships when it ran aground near the mouth of the Yangtze River.

Another theory is that the Peking Man fossils are still hidden in China. Newspaper articles over the years have made various suggestions as to who might be hiding them there. Harry Shapiro quoted Pei Wenzhong as saying that not long after their disappearance, bones that looked like the missing relics were found on the Chinese coast, about to be loaded on another ship (but the remains were later proven not to be the Peking Man fossils). Pei noted that from then on, the Japanese were no longer interested in finding the fossils, and he theorized that they had found the remains in the port and transported them to Japan.

This same notion is reflected in the story of Frank Whitmore, Jr., a geologist employed by the U.S. Army in Tokyo, who wrote to Harvard professor Tilly Edinger on November 8, 1945, two months after the Japanese surrender. Whitmore claimed that he had discovered the Peking Man fossils in a collection at Tokyo University, together with Davidson Black's original records, and that he was making arrangements to have the fossils sent back to Peking Union Medical College. Two weeks later, Whitmore wrote Edinger with a more exact description of the finds. According to Harry Shapiro, Whitmore's description fit that of the remaining small items left at Peking Union Medical College—not the actual fossils of Peking Man. The fact that the Japanese would take away even such a minor set of fossils, Shapiro noted, shows that the Chinese and Americans were right in fearing for the fate of the original Peking Man finds.

According to other reports, the crates of fossils were loaded onto the train taking the Marines to the Chinese coast to board a ship heading for the United States. The train was stopped by the Japanese, who arrested the Marines, and the Peking Man containers disappeared. One theory proposes that the Japanese, assuming nothing of interest, didn't even open the crates, and threw them out of the ransacked train; thus the fossils are lost but might still be found somewhere in China. Another version of this story has the Japanese taking the boxes off the train and shipping them to Japan.

In Pei's view, the fossils indeed made it to Japan and are still there. Others believe that after the Japanese surrender, U.S. military personnel in Japan repossessed the Peking Man fossils, but instead of sending them back to China took them to the United States, where they are now hidden.

Yet another theory holds that the Nationalist Kuomintang army was in the area when the train carrying the Marines was stopped by the Japanese. The Nationalists took possession of the crates and transported them to Taiwan with other booty, as Chiang Kai-shek and his forces fled the mainland.

Harry Shapiro recounted an interesting story about D. M. Watson, an English paleontologist who visited the American Museum of Natural History. Franz Weidenreich, who still held a position in anthropology at the time, showed Watson casts of the Peking Man fossils he had brought to New York at the beginning of the war. Back in England, Watson told his students what he had seen in America, stressing that these were casts. One of his students, a German with purported communist leanings, according to Shapiro, either misunderstood or purposely withheld this fact when he later visited China and recounted to Communist Chinese authorities what Watson had seen. There followed an official Chinese complaint to the United States, and newspaper reports that the Peking Man fossils were actually in New York. These included an article in *The New York Times*, to which Shapiro responded in a published letter denying the claim.

In February 2006, I gave a lecture at the American Museum of Natural History, and used the opportunity to visit Ian Tattersall, the curator of anthropology at the museum. He opened the drawers behind his desk and showed me the casts of the Peking Man fossils that Franz Weidenreich had brought to the museum. I was amazed by their high quality, and

I could see how someone not trained in paleontology might think these were the real fossils.

Harry Shapiro and Claire Taschdjian spent enormous amounts of time and effort researching what might have happened to Peking Man, and each has written a book about this mystery. When it became publicly known in the early 1970s that Shapiro was keen on solving this enigma, a number of people contacted him, and news reports appeared about leads in the case. In April 1971, Shapiro received a call from someone speaking on behalf of Dr. William Foley, a career Marine Corps officer who had served with the contingent that was to be evacuated from China when the United States and Japan went to war. Foley was interested in helping Shapiro because he was aware that he had been wrongfully accused by the Chinese of harboring allegedly stolen fossils.

The two men met, and Foley recounted that in 1941 he was stationed at Camp Holcomb in Qinhuangdao, up the coast and northeast of Tientsin. He was in charge of a medical unit of seventeen Marines. As the political situation deteriorated, the decision was made to evacuate the Marines to the Philippines aboard the *President Harrison*. At about that time, two footlockers were brought to Camp Holcomb, marked with Foley's name and tagged as "Personal Possessions," and he was told to take care of them until the contingent's shipment out of China, scheduled for December 8. This was December 7 in Hawaii, and the attack on Pearl Harbor had just taken place. Japanese soldiers immediately surrounded the Marines in Camp Holcomb, who were intending to resist them. The Marines were rushing to arms when a call came through from high command ordering them to surrender— there was no hope of defending themselves against the far more numerous Japanese forces.

The Marines were taken to a temporary POW camp in Tientsin, along with their belongings, including the two footlockers. Since he was a ranking officer, the Japanese gave him some freedom, and for about a week he was allowed to go to Peking and move at will. He told Shapiro that he saw the Peking Man skulls packed in glass jars rather than wooden crates—which was the first inconsistency with Claire Taschdjian's report of the packing operation.

Foley recounted that, once back at the POW camp, he opened the footlockers and distributed their contents between the Swiss warehouse and the Pasteur Institute, both in Tientsin, in order to protect them. Another inconsistency Shapiro noted was Foley's claim now that there were four footlockers. One footlocker remained with Foley, who believed it contained the most important relics.

Foley spent the next four years as a prisoner of war, and was transferred between POW camps three times. He said that he was able to keep the Japanese from opening his footlocker in all four camps—a claim Shapiro found hard to believe. Just before liberation in 1945, Foley was separated from his footlocker, and he never saw it again.

Shapiro became intrigued by the possibility that U.S. personnel returning from China at war's end, especially Marines who had been stationed there before Pearl Harbor, might have some information on the missing fossils—and perhaps possess some of the bones. He wanted the Chinese to be aware of the efforts made in the United States to retrieve the missing treasure. Relations were warming up between the two nations, and President Nixon's historic visit would come around this time, February 1972.

Shapiro's article about his search for the fossils had appeared in the journal of the Museum of Natural History, and he made sure that one of the aides accompanying National Security Advisor Henry Kissinger to China before Nixon's official visit had copies of the article to present to Chinese officials. Shapiro's article and his quest for Peking Man were widely publicized; there had been articles in *The New York Times* and other papers. He had hoped that Chinese paleontologists would want to follow up on his leads from Dr. Foley's story. But he heard from no one.

In the United States interest in this matter continued, and Shapiro received many letters from the public. One correspondent was Christopher Janus, a Chicago businessman who said he had become fascinated by the disappearance of Peking Man and decided to launch his own investigation. Janus had advertised in newspapers for possible leads from U.S. personnel who had been stationed in China during the war, offering a reward for any useful information.

He received a reply from a woman who told Janus that her late husband had served with the Marines in China, and that before he died he had entrusted her with a box containing bones he said were extremely valuable. He had warned her that she might be in danger if information about this bone collection was revealed.

The woman was evidently frightened, but she also wanted the reward money; in fact, she wanted much more than the advertised amount. After lengthy phone discussions to arrange when and where to meet, Janus and the woman decided on a day and hour, and the top of the Empire State Building. They identified each other by agreed-upon signs. The woman showed Janus a picture of a box containing bones and one flat skull. But suddenly, when she saw a tourist raise a camera to take a picture of the view—this was, after all, the top of the Empire State Building—she became terrified, ran to the elevator, and disappeared.

Janus consulted with Shapiro about what to do next. They decided that he should place an ad in the *Times*—guessing that the woman had chosen the top of the Empire State Building because she lived in or near New York. Some weeks later, she did indeed call Janus. He tried to persuade her to have Harry Shapiro authenticate the bones.

After long negotiations, the woman gave a picture of the bones to Janus, who forwarded it to Shapiro. Even though most of the bones seemed new, or at least not from the Peking Man collection, the skull in the photograph looked "Sinanthropine," as Shapiro put it. Later, at an international anthropology meeting in Chicago to which Janus took

the photograph, Professor William Howells of Har~~~~~~~~~~~~~~~
was inclined to identify the skull in the photograph as Skull XI in the
Weidenreich list. Other anthropologists were cautiously optimistic
that the single skull in the box could have been that of a *Homo erectus*,
and possibly from the lost Peking Man collection.

But the negotiations with the woman failed, and the trail went
cold. In an effort to locate this reluctant person, or someone else who
might have information on the missing fossils, Shapiro researched the
identities of all Marines who had been evacuated from Camp Holcomb
and had left behind widows who now lived in the New York area. He
discovered one woman who fit the description of such a widow, then
enlisted the FBI to help find her. In January 1973, the FBI came up with
the identity of the only widow of a Camp Holcomb Marine living in
the New York area, in New Jersey. But this was not the mystery
woman. The FBI continued its investigation, to no avail.

In his own book on the puzzle of the missing Peking Man fossils,
Christopher Janus went further in relating this curious episode: Sometime after the trail of "The Empire State Building Lady," as he called her,
went cold, he received a call from a man who identified himself as Harrison Seng and claimed to be the woman's attorney. In exchange for the
bones, he demanded from Janus $500,000 (later raised to $750,000)
and a letter from the People's Republic of China stating that the Chinese government would never charge his client with wrongdoing for
keeping the bones. Janus could not provide the desired Chinese legal
promise, and the discussion with Seng broke off.

Janus has said that he spent much of the 1970s looking for Peking
Man. He traveled to China, Taiwan, and Hong Kong in search of clues
about the missing fossils. He became convinced that despite the fact
that the Nationalist Chinese had given the order to evacuate the fossils
to America, and that they subsequently took much art and many other
treasures with them to Taiwan, the missing fossils were not likely to be
found there.

Another lead Janus followed was the story that the crates with the fossils were taken to the Soviet Union. A Russian-American man named Alexis Petrov, whom Janus met in New York in 1973, told him that he had lived in Shanghai during the war, teaching high school biology, and that he knew something about Peking Man. He said that the fossils had been loaded onto a Russian ship and taken to Yalta. But with no supporting evidence for this claim, Janus thought that it was probably false.

This leaves only a few valid theories, out of the many that have been proposed over the years, as to the fate of Peking Man. One is that Camp Holcomb Marines or their families or descendants still have at least some of the fossils—presumably hidden from the Japanese during the long occupation of China and the Marines' incarceration in POW camps. Another theory, one that Janus was not able to pursue, is that the fossils are hidden in Japan.

A third viable theory holds that the fossils were sold by Japanese soldiers to Chinese peasants, who ground them up to make folk remedies; thus the "dragon bones" ended up fulfilling their usual purpose in China. While Japanese authorities were interested in the fossils, soldiers in the field might not have been aware of this quest, and all they cared about, according to some American Marines, was looting and pilfering. If they found the crates at the camp, in a railroad car, or elsewhere, the Japanese soldiers likely either threw their contents away or tried to sell them to Chinese peasants.

According to Janus, Teilhard de Chardin maintained that the fossils might have been dumped, by mistake or intentionally, into the Yellow Sea. Teilhard was very familiar with Tientsin, and had the fossils been left there, perhaps he could have helped find them.

On February 26, 1981, *The New York Times* reported that Christopher Janus had been indicted on thirty-seven counts of fraud for funneling $640,000, ostensibly raised for searching for the lost Peking Man fossils, to personal use. This cast a deep shadow over Janus and his work, and ended a perhaps less than genuine attempt to use private funding to locate the lost treasure.

But not all hope had been lost. In the fall of 1980, Harry Shapiro, who remained committed to finding any information he might obtain about the disappearance of the relics, traveled to China with his daughter. There he met Chinese colleagues, and with them visited the Pasteur Institute and the Swiss warehouse in Tientsin—locations that might have housed the fossils—but found no trace of the lost crates or their contents. He then traveled to the site of Camp Holcomb. He had been told that the place to look for the remains was beneath the wooden floor of the cellar of Building 6 of the former U.S. Marine barracks. Shapiro had with him a picture of the building taken before the war, and within a short time the exact location was identified. With the help of local authorities, he searched the premises extensively. But then he learned that four years earlier, in 1976, the old barracks had collapsed in an earthquake; it had since been rebuilt as a school of public health. Shapiro and his Chinese colleagues found nothing.

In March 1980, before Shapiro's visit, Jia Lanpo had met the Swiss ambassador to China, who was interested in archaeology and seemed to know much about the search for Peking Man. He suggested to Jia that the Chinese might do well to try to locate William Foley as a source of information on this mystery. A friend of the former Marine told Jia that Foley, who was working as a doctor in Tientsin,

was supposed to return to the United States on a ship (this must have been in 1941) when his friend and neighbor Pierre Teilhard de Chardin, afraid of the advance of Japanese troops, asked him to shelter in his house the cases containing the famous fossils. Before Dr. Foley was able to move these cases to the ship, together with his own belongings, he had himself to flee in a hurry, leaving everything behind in his house, which was immediately occupied by Japanese army men. According to Dr. Foley, these Japanese had been the last to see the valuable cases with the fossils, and his guess is that the cases were shipped to Japan later on, when the house was totally emptied by the Japanese. His further guess is that the cases arrived in Japan—providing the carrying ship was not sunk.

This account contradicted what Foley had told Shapiro. Foley was contacted and invited to China to help search for the lost fossils, but for political reasons, he could not go. His wartime house was identified by Chinese investigators, who entered it to search for the missing "cases." Nothing was found.

The search for Peking Man continued. In February 2000, reports in the Chinese media, principally the journal *Guangming Ribao*, claimed that a former Japanese soldier, Nakada Hironami, had testified that in January 1945 he saw a skull that he now believed was part of the Peking Man collection in the possession of his father-in-law, who had been a high official in the puppet government the Japanese had set up in Manchuria. Hironami claimed that he later saw the same skull at his father-in-law's house in Japan, after the Japanese surrendered and retreated from China. Nothing more is known about this story.

Testimony from two former employees at the U.S. embassy in Peking gave Chinese authorities another tip. The Americans said that in 1941 they saw a crate they now thought may have contained the Peking Man fossils being buried at the back of the embassy. Chinese investigators found the location indicated by the Americans; it was now the floor of a garage. But they did not dig up this floor to look for the crate.

In September 2005, the Chinese government announced that it would spare no effort to locate the missing fossils—every tip would be followed. Surprisingly, officials declared that they did not believe the fossils were in the United States, Russia, or Taiwan—or even at the bottom of the sea. Instead they felt that the fossils were in either China or Japan. Liu Yajun, the government's chief investigator, stated in 2006 that since its establishment the previous July, the search committee had received sixty-three credible tips, in addition to twenty-one earlier tips that had been sent to the UNESCO World Heritage center at Zhoukoudian. These ranged from sightings of Peking Man skulls in Japan to names of people—including one who claimed to be 121 years old—with supposed knowledge about the missing fossils. To date, nothing substantial has surfaced. It should be noted that there has been no report of a serious effort to find the missing fossils in Japan.

TEILHARD HIMSELF was interrogated in 1941 by the Japanese, who thought he might know something about the whereabouts of Peking Man. A decade later, in the 1950s, he wrote these lines to a friend:

> The famous skullcap—might it be hidden in Japan? Wouldn't it have been destroyed, perhaps, by an ignorant plunderer? Might it be buried in a garden in Peking? Perhaps we might retrieve it someday, in its million-year freshness, from some courtyard in Peking?

Like everyone else involved, Teilhard was deeply upset about the loss of the fossils he had worked hard to help discover and analyze. These lost relics represented a major part of the tangible results of his life's work in science, and stood as a symbol in his philosophy of human evolution and our eternal search for origins.

Pierre Teilhard de Chardin's work on Peking Man demonstrates that a common ground between science and faith is something we may still strive for. It is akin, perhaps, to our hope that the lost fossils of Peking Man may someday be found. Even without the recovery of the fossils, science progresses and our knowledge of the universe and our place in it expands continually. The story of our species, from its humble origins on the African plains through *Homo erectus* in Asia, to the Neanderthals

Teilhard de Chardin in New York, 1950. Teilhard de Chardin collections, Georgetown University.

and Cro-Magnons of Europe, is explored ever more deeply, and its details are constantly being uncovered.

Teilhard's wholehearted embrace of evolution attests to the enduring nature of Darwin's theory. The fact that a devout Jesuit never wavered in his espousal of this theory speaks volumes for its ability to explain the nature of life and our experience on Earth.

His posthumously published book *The Future of Man* begins with echoes of Galileo, as Teilhard sets forth his view of science, religion, and philosophy:

The conflict dates from the day when one man, flying in the face of appearance, perceived that the forces of nature are no more unalterably fixed in their orbits than the stars themselves, but that their serene arrangement around us depicts the flow of a tremendous tide—the day on which a first voice rang out, crying to Mankind peacefully slumbering on the raft of Earth, "We are moving! We are going forward!"

Teilhard understood the gravity and the depth of the implications of our struggle to combine science and faith. He knew how hard it is to achieve what he had started: a wedding of science with belief, and an embrace of the theory of evolution by everyone, scientists, the religious, the public in general. He knew that in science and faith, the road ahead is long and arduous. And he understood that working alone, he could not bring all people to one goal. Others had to contribute to this important effort. Late in his life, Teilhard wrote:

If I have had a mission to fulfill, it will only be possible to judge whether I have accomplished it by the extent to which others go beyond me.

ACKNOWLEDGMENTS

I am thankful to my sister, Ilana Aczel, who in the summer of 2005 urged me to join her on her fourth trip to China. It was during my visit to Beijing that I first developed the idea for this book. I am equally thankful to my late father, E. L. Aczel, who introduced me to the writings of Pierre Teilhard de Chardin, which, thirty years later, inspired me to pursue this story.

Numerous researchers, academics, and religious experts kindly offered me their help in following this tale of human evolution, paleoanthropology, and the relation between science and faith. They include Professor Russell L. Ciochon of the Department of Anthropology at the University of Iowa; Professor Dong Wei of the Institute of Vertebrate Paleontology and Paleoanthropology in Beijing; Dr. Ian Tattersall, Curator of Anthropology at the American Museum of Natural History in New York; Professor Ofer Bar-Yosef, MacCurdy Professor of Prehistoric Archaeology and Curator of Paleolithic Archaeology at Harvard University; Professor David Pilbeam of the Department of Anthropology at Harvard; and Father Thomas King, S.J., of the Department of Theology at Georgetown University. It must be noted that the views on evolution, science, and religion expressed in this book do not necessarily reflect those of any of these individuals.

I am grateful to the librarians at Harvard University's Widener

Library and at Boston University's Mugar Library for access to many important sources; and to the administrator and staff of the Fondation Teilhard de Chardin in Paris, the repository of many documents and photographs related to the life of Teilhard, and to the staff of the Bibliothèque Nationale de France, also in Paris, for assistance in my research. I acknowledge the administrators of the Institute Library and Roman Archives of the Society of Jesus in Rome. I thank the administration of the Teilhard de Chardin collections at Georgetown University for access to a number of photographs. I am indebted to the Center for the Philosophy and History of Science at Boston University, and to its director, Professor Alfred Tauber, for my research fellowship at the Center; and to Professor Owen Gingerich and the Department of the History of Science at Harvard University for my position of visiting scholar.

I am most grateful to my agent, John Taylor ("Ike") Williams, of Kneerim & Williams in Boston, for his support, guidance, and many ideas in preparing this book. Also at Kneerim & Williams, I warmly thank Brettne Bloom for her tireless efforts to turn a mere concept into a viable project, and for her many hours of work on the book proposal. My gratitude goes as well to Hope Denekamp at Kneerim & Williams for her enthusiastic assistance.

I thank the publisher of Riverhead Books, Geoffrey Kloske, for his ideas and encouragement. Jake Morrissey is by far the best editor any author could hope for. Thank you, Jake, for everything you have done for this book; it is always a pleasure to work with you. I am indebted to Anna Jardine for her incomparable copyediting of the manuscript, which improved it immensely. Many thanks also to Sarah Bowlin for her help with the book, and to Sarah Walsh for her work on the design.

Finally, I thank my wife, Debra, for her many suggestions, for her photographs, and for her wise advice.

Appendix 1

The following chart shows Paleolithic, or Old Stone Age, cultures and their approximate dates, in years before the present. The chart was created by Professor Ofer Bar-Yosef of Harvard University.

Geological Periods	PALEOLITHIC	Western Europe	Africa	Western Asia	Eastern Asia	Australia
11,500	Upper Paleolithic 45,000–35,000	Magdalenian Solutrean Gravettian Aurignacian Châtelperronian 40,000	Late Stone Age (Nile Valley) 45,000	45,000	Zhoukoudian upper cave	Arrival of humans 50,000
PLEISTOCENE	Middle Paleolithic 250,000	Mousterian cultures	Middle Stone Age	Mousterian cultures		
	Lower Paleolithic	Acheulean complex 500,000	Acheulean complex	Acheulean complex	Zhoukoudian Layer 3	
				1,400,000 Core and Flake industry	Zhoukoudian Layer 10 600,000	
			1,700,000 Oldowan		Core and Flake industry	
1,600,000 PLIOCENE	2,500,000					

Appendix 2

RADIOCARBON AND OTHER
SCIENTIFIC DATING METHODS

In 1946, the American chemist Willard Libby published a paper suggesting that living organisms may contain radioactive forms of carbon. A year later, together with Ernest Anderson and other colleagues, he published a paper in the journal *Science* that described the detection of radioactive carbon (radiocarbon) in biological material.

Living systems absorb carbon throughout their lifetimes. This carbon is, in a small proportion, the radioactive form carbon 14 ("normal" carbon is carbon 12; there is also an isotope with weight 13, which is corrected for in calculations). Carbon 14 is formed constantly in the upper atmosphere because of the bombardment of nitrogen atoms by radiation from the sun. Living things are at equilibrium with their environment, meaning that the proportion of radioactive carbon in them is the same as in the environment. But when an organism dies, its radioactive carbon decays and thus its proportion to the rest of the carbon in the dead organism diminishes. Knowing that the half-life of carbon 14 is 5,730 years (Libby assumed that it was 5,568 ± 30 years) allows us to estimate the time that has elapsed since the organism died and stopped absorbing carbon 14.

Radiocarbon dating has been perfected over the decades, and its accuracy has been calibrated against very reliable dating systems such as counting tree rings. A tree adds a ring a year; establishing a time frame doesn't get any better than that in terms of accuracy. Radiocarbon dating, which works for up to around 50,000 years, can be used to date some Neanderthal and anatomically modern human sites. For earlier dates, similar radioactive-decay methods are used, applied to rocks rather than remains of living organisms. These techniques involve uranium or potassium-argon.

NOTES

Complete citations for works cited here in brief may be found in the References section.

Abbreviations:

MB	Marcellin Boule
HB	Henri Breuil
LS	Lucile Swan
TdC	Pierre Teilhard de Chardin
MTC	Marguerite Teillard-Chambon

CHAPTER 1 · THE BANQUET

13 **gold-sashed Chinese servants:** Lukas and Lukas, *Teilhard*, pp. 101–102.

16 **was named after him:** Boaz and Ciochon, *Dragon Bone Hill*, p. 9.

16 **prehistoric tools, as we now know:** Ibid., p. 95.

19 **". . . expeditions to mythological caves":** Reported in Walker and Shipman, *The Wisdom of the Bones*, p. 57.

20 **". . . the Central Asiatic origin of the Hominidae":** Jia and Huang, *The Story of Peking Man*, p. 27.

20 **". . . Even his manner had an unusual charm":** Shapiro, *Peking Man*, p. 40.

CHAPTER 2 · PRELUDE TO EVOLUTION

28 **Foucault's definitive proof of Earth's rotation:** For a survey of the development of these ideas, see Aczel, *Pendulum*.

29 **fossils were the remnants of living things:** Trinkaus and Shipman, *The Neandertals*, pp. 14–15.

30 **of each living creature within the universe:** Ibid., p. 19.

CHAPTER 3 · DARWIN'S BREAKTHROUGH

33 **seemed a ridiculous idea to most people:** Trinkaus and Shipman, *The Neandertals*, p. 10.

39 **". . . two or three years had elapsed":** Quoted in Richard E. Leakey, *The Making of Mankind*, p. 25.

40 **the Linnean Society in London:** Ibid., p. 26.

40 **". . . and at rare intervals":** Darwin, *The Origin of Species*, p. 382.

40 **". . . which has had a history":** Ibid., p. 381.

40 **within our relationship to the past:** George Levine, introduction to Darwin, *The Origin of Species*, p. xv.

41 **". . . but success in leaving progeny":** Darwin, ibid., p. 61.

41 **this process could be explained:** Kennedy, *Paleoanthropology*, p. 1.

42 **". . . took away our breath":** Landau, *Narratives of Human Evolution*, p. 1.

42 **and pandemonium ensued:** Reported in pamphlet, of Natural History Museum, London.

43 **". . . jumped out of my seat":** Quoted in Leonard Huxley, *The Life and Letters of Thomas Henry Huxley*, p. 194.

43 **represent the "missing link":** Walker and Shipman, *The Wisdom of the Bones,* p. 32.

CHAPTER 4 · STONE TOOLS AND CAVE ART

45 **near Hoxne, in Suffolk:** Trinkaus and Shipman, *The Neandertals*, p. 33.

45 **"had not the use of metals":** Frere, "An Account of Flint Weapons Discovered at Hoxne in Suffolk."

46 **Neandertal or Neanderthal in German:** While the spelling of the German word for "valley," *Thal,* has been modernized to *Tal,* most people

use the older spelling for "Neanderthal." Some scientists, however, follow the new German orthography, thus "Neandertal." I see no reason for doing so.

48 **the harbor at Haifa:** Ofer Bar-Yosef, personal communication, October 20, 2006. See also Callander and Bar-Yosef, "Saving Mount Carmel Caves: A Cautionary Tale for Archaeology in Our Times."

48 **symbolic remembrance of the dead:** Trinkaus and Shipman, *The Neandertals*, p. 339.

49 **much stronger than ours:** Johanson, Johanson, and Edgar, *Ancestors*, p. 261.

49 **hundreds of thousands of stone tools have been found:** Stringer and Gamble, *In Search of the Neanderthals*, p. 15.

49 **18,000 to 11,000 years ago:** Here and elsewhere, dates are approximate; different experts often use different dates for the same period. For example, the range 11,000 to 18,000 years before the present is also used by some archaeologists for the Magdalenian age.

56 **Pioneer Man:** Christopher Stringer, "The First Humans North of the Alps," *British Archaeology* 86 (January/February 2006).

56 **direct ancestors of the Neanderthals:** Johanson, Johanson, and Edgar, *Ancestors*, p. 260.

57 **may have possessed language:** Arensburg et al., "A Middle Paleolithic Human Hyoid Bone," pp. 758–60.

58 **"... around the campfire in their caves":** Johanson, Johanson, and Edgar, *Ancestors*, p. 264. Johanson prefers the spelling "Neandertal."

59 **anatomically modern humans occupied Europe:** Bar-Yosef, "The Role of Western Asia in Modern Human Origins."

59 **a thousand or two thousand years:** Mellars, "A New Radiocarbon Revolution and the Dispersal of Modern Humans in Eurasia," p. 933.

60 **for long-term survival:** Ofer Bar-Yosef, "Moving Out: Human Migration in Early Prehistory," lecture given at Boston University, October 17, 2006.

60 **in making this determination:** Stringer and Gamble, *In Search of the Neanderthals*, p. 16.

60 **"rudimentary intellectual faculties":** Ibid., p. 25 (from Boule and Vallois, *Fossil Men*).

61 **Neanderthals possessed a language:** Dumiak, "The Neanderthal Code," pp. 22–25.

CHAPTER 5 · JAVA MAN

63   **present-day Indonesia:** This chapter draws on material in Shipman, *The Man Who Found the Missing Link.*

67   **headed for the East:** Ibid., p. 76.

69   **". . . those who want to see his fossils":** Ibid., p. 343.

69   **". . . most maligned fossil" ever found:** Johanson, Johanson, and Edgar, *Ancestors*, p. 181.

CHAPTER 6 · TEILHARD

71   **volcanoes to steep ravines:** Speaight, *The Life of Teilhard de Chardin*, p. 25.

72   **Astorg Teilhard was also ennobled:** Cuénot, *Teilhard de Chardin*, p. 1.

73   **the rest of the students awoke:** Lukas and Lukas, *Teilhard*, p. 25.

73   **he would have loved to do:** Speaight, *The Life of Teilhard de Chardin*, p. 29.

73   **". . . when it is being persecuted":** TdC, March 25, 1901, in Teilhard de Chardin, *Images et Paroles*, p. 25. Author's translation.

75   **he saw in this All:** King, *Teilhard's Mass*, p. 1.

78   **". . . an absolute contempt for danger":** Speaight, *The Life of Teilhard de Chardin*, p. 59.

78   **". . . to recover a wounded soldier":** Ibid., p. 64.

78   **". . . of which we have no knowledge":** Teilhard de Chardin, "The Cosmic Life," in *Writings in Time of War*, p. 14.

79   **". . . with the life-blood of evolution":** Ibid., p. 57.

79   **". . . for our peaceful readers":** Arnould, *Teilhard de Chardin*, p. 196. Author's translation.

79   **". . . as manuscripts passed under a coat":** TdC to MTC, December 23, 1916. Fondation Teilhard de Chardin, Paris. Author's translation.

80   **the right of the Church:** Speaight, *The Life of Teilhard de Chardin*, p. 75.

80   **"a naturally pantheistic soul":** Ibid., p. 80.

80   **". . . the Powers of the Earth":** Teilhard de Chardin, "Le Prêtre" (July 8, 1918), in *Images et Paroles*, p. 58. Author's translation.

82   **". . . hardships he constantly shared":** Speaight, *The Life of Teilhard de Chardin*, p. 121.

83   **"with highest honors":** Lukas and Lukas, *Teilhard*, p. 71.

83   **cruelty we see in this world:** Ibid., p. 72.

85 **Aristotelian-Thomist doctrine of the Church:** Ibid., p. 74.

86 **"Arrive 15 May":** Arnould, *Teilhard de Chardin*, p. 103. Author's translation.

86 **". . . to escape this inner tearing":** Lukas and Lukas, *Teilhard*, p. 75.

87 **". . . the comforts of life":** Arnould, *Teilhard de Chardin*, p. 104. Author's translation.

88 **". . . the voice of the Burning Bush":** Speaight, *The Life of Teilhard de Chardin*, p. 123.

CHAPTER 7 · A DISCOVERY IN INNER MONGOLIA

89 **the lost Garden of Eden:** Jia and Huang, *The Story of Peking Man*, p. 18.

90 **had no idealism or hope:** Lukas and Lukas, *Teilhard*, p. 79.

91 **deserved a more thorough exploration:** Jia and Huang, *The Story of Peking Man*, p. 18.

92 **discovered the Shuidonggou site:** Ibid., pp. 18–19.

93 **a meal of eggs and potatoes:** Ibid., p. 19.

93 **". . . labors and sufferings of the world":** Teilhard de Chardin, "The Mass on the World," as translated in King, *Teilhard's Mass*, p. 145.

93 **". . . other forms of human activity":** Quoted in Speaight, *The Life of Teilhard de Chardin*, p. 129.

94 **". . . It's the real free life!":** TdC to HB, August 14, 1923, Teilhard de Chardin, *Letters from a Traveller*, pp. 83–84.

95 **". . . twenty Mongols and Chinese":** TdC to Abbé Gaudefroy, August 15, 1923, Teilhard de Chardin, *Lettres à l'Abbé Gaudefroy et l'Abbé Breuil*, pp. 24–26. Author's translation.

96 **". . . bone clearly worked on":** TdC to HB, July 16, 1923, Teilhard de Chardin, *Lettres à l'Abbé Gaudefroy et l'Abbé Breuil*, pp. 130–132. Author's translation.

96 **". . . I must proceed methodically":** TdC to HB, July 25, 1923, ibid., p. 136. Author's translation.

96 **". . . admirable 'Mousterian' points":** TdC to MB, September 29, 1923, Teilhard de Chardin, *Teilhard de Chardin en Chine*, p. 75. Author's translation.

98 **date it to the Pleistocene:** Jia and Huang, *The Story of Peking Man*, p. 22.

100 **". . . last letter I will write you from China":** Teilhard de Chardin, *Letters to Léontine Zanta*, pp. 67–68.

101 **". . . intrepid ones, not the innocents":** Cuénot, *Teilhard de Chardin* (1958), p. 82. Author's translation.

102 **Vladimir Ledochowski:** Speaight, *The Life of Teilhard de Chardin*, p. 136.

102 **the most difficult to accept:** Ibid., p. 137.

102 **". . . they want me to sign":** TdC to Auguste Valensin, November 13, 1924. Fondation Teilhard de Chardin, Paris. Author's translation.

103 **". . . to be excommunicate":** Lukas and Lukas, *Teilhard*, p. 104.

104 **the exact content of these secret documents:** Thomas M. King, S.J., personal communication to author, September 6, 2006.

CHAPTER 8 · *AUSTRALOPITHECUS* AND
THE SCOPES TRIAL

106 **his fossil and the new find:** Trinkaus and Shipman, *The Neandertals*, pp. 224–226.

107 **permanent first molar just emerging:** Johanson, Johanson, and Edgar, *Ancestors*, p. 142.

108 **"in adoration of our ancestor":** Trinkaus and Shipman, *The Neandertals*, p. 233.

109 **whose discovery was announced in 2006:** John Noble Wilford, "Little Girl, 3 Million Years Old, Offers New Hints on Evolution," *The New York Times*, September 21, 2006, pp. A1, A10.

112 **reported the ACLU's plans:** Tompkins, *D-Day at Dayton*, pp. 11–12.

112 **permanent damage to his career:** Ibid., p. 12.

113 **". . . freedom would have lost":** Scopes and Presley, *Center of the Storm*, p. 4.

113 **to lead an opening prayer:** Levy, *The World's Most Famous Court Trial*, p. 3.

114 **". . . known as the anti-evolution statute":** Ibid., p. 4.

114 **". . . male and female created He them":** As transcribed in the court record quoted ibid., p. 6.

116 **". . . powers of the State of Tennessee":** Quoted in Tompkins, *D-Day at Dayton*, p. 37.

116 **". . . outrage the jury with their blasphemies":** Quoted ibid., pp. 38–44.

116 **". . . and we are not afraid of it":** Quoted ibid., p. 65.

117 **". . . both to God and to human intelligence":** As transcribed in the court record quoted in Levy, *The World's Most Famous Court Trial*, p. 252.

117 **". . . without using the theory of evolution":** As transcribed in the court record quoted ibid., p. 241.

117 **". . . without teaching evolution":** As transcribed in the court record quoted ibid., p. 238.

118 **". . . I think the fine is unjust":** Quoted in Tompkins, *D-Day at Dayton*, p. 53.

118 **". . . before the Hun is at their gates":** Quoted ibid., p. 51.

CHAPTER 9 · THE EXILE

122 **return to China and rejoin Licent:** Arnould, *Teilhard de Chardin,* p. 144.

123 **". . . too much in the public view":** TdC to Auguste Valensin, May 19, 1925. Fondation Teilhard de Chardin, Paris. Author's translation.

123 **could never live outside this science:** TdC to Ida Treat, April 6, 1927, Arnould, *Teilhard de Chardin*, p. 145.

125 **". . . That which I love I see no more":** TdC to MTC, April 26, 1926, Teilhard de Chardin, *Lettres de Voyage (1923–1955)*, pp. 130–131. Author's translation.

125 **"if only I were ten years younger":** TdC to MTC, April 26, 1926, ibid., p. 133. Author's translation.

126 **"The Pirate of the Red Sea":** Arnould, *Teilhard de Chardin,* p. 166.

126 **". . . doesn't interest me at all":** Grandclément, *L'Incroyable Henry de Monfreid*, p. 275. Author's translation.

127 **". . . animals (buffalo, elephants, tigers) disappear":** TdC to MTC, April 26, 1926, Teilhard de Chardin, *Lettres de Voyage (1923–1955)*, p. 133. Author's translation.

128 **". . . without any communication for several weeks":** TdC to MTC, July 8, 1926, ibid., p. 136. Author's translation.

128 **". . . without much hindrance":** Ibid., pp. 136–137. Author's translation.

128 **". . . to reach Gansu this year":** TdC to MB, July 19, 1926, Teilhard de Chardin, *Teilhard de Chardin en Chine*, p. 145.

129 **". . . nothing recognizably human":** TdC to Abbé Gaudefroy, October 12, 1926, Teilhard de Chardin, *Lettres à l'Abbé Gaudefroy et l'Abbé Breuil*, p. 51.

131 **Rome, where it was examined:** TdC to Joseph Lévie, October 23, 1932, as reported in Speaight, *The Life of Teilhard de Chardin*, p. 147.

131 **to continue his scientific work:** Lukas and Lukas, *Teilhard*, p. 106.

132 **gave a copy to the Abbé Bruno de Solages:** The title of the paper in French is "Le Fondement et le Fond de l'Idée d'Évolution." Arnould, *Teilhard de Chardin*, p. 165.

132 **interviewed him at length about this subject:** Speaight, *The Life of Teilhard de Chardin*, p. 156.

134 **pain of his many stings without a word:** Arnould, *Teilhard de Chardin*, p. 170.

134 **". . . awaiting us with open arms!":** TdC to MB, November 26, 1928, Teilhard de Chardin, *Teilhard de Chardin en Chine*, p. 178. Author's translation.

135 **wrote to Marguerite . . . visiting the Afar region:** TdC to MTC, December 28, 1928, Teilhard de Chardin, *Lettres de Voyage (1923–1955)*, pp. 158–159. Author's translation.

135 **pick up a shipment of hashish:** Arnould, *Teilhard de Chardin*, pp. 153-154.

135 **". . . in order to save the World":** Quoted in Lukas and Lukas, *Teilhard*, p. 109.

CHAPTER 10 • THE DISCOVERY OF PEKING MAN

137 **fossils . . . would be kept and analyzed:** Walker and Shipman, *The Wisdom of the Bones*, p. 58.

138 **caravanserai called the Liu Shen Inn:** Ibid., p. 58.

138 **more money—and got it:** Ibid., p. 59.

140 **". . . start with an article in *Nature*":** TdC to MB, March 24, 1929, Teilhard de Chardin, *Teilhard de Chardin en Chine*, p. 189. Author's translation.

141 **". . . very bottom of the fossiliferous deposits":** Jia and Huang, *The Story of Peking Man*, p. 61.

141 **pig and buffalo skulls, and deer antlers:** Ibid., p. 62.

143 **decided to continue to explore it:** Ibid., p. 63.

143 **"a corking field man":** Walker and Shipman, *The Wisdom of the Bones*, p. 59.

144 **". . . and work with the other":** Jia and Huang, *The Story of Peking Man*, p. 64.

145 **"Found skullcap . . . like man's":** Ibid., p. 65.

145 **". . . turns out to be true":** Ibid., pp. 65–66.

146 **". . . to arrest me first":** Ibid., pp. 67–68.

147 **". . . as one could wish for":** Lukas and Lukas, *Teilhard*, 1977, p. 113.

148 **". . . has barely been touched":** TdC to MB, December 11, 1929, Teilhard de Chardin, *Teilhard de Chardin en Chine*, p. 190. Author's translation.

150 **to about 410,000 years ago:** Boaz and Ciochon, *Dragon Bone Hill*, p. 116.

150 **now sprawling excavation complex:** Walker and Shipman, *The Wisdom of the Bones*, p. 63.

CHAPTER 11 • TEILHARD MEETS LUCILE SWAN

152 **". . . one answers with a blow":** Quoted in Lukas and Lukas, *Teilhard*, p. 114.

152 **". . . beyond the common measure":** Pierre Leroy, S.J., "The Man," in Teilhard de Chardin, *Letters from a Traveller*, p. 15.

154 **there was a God:** Mary Wood Gilbert, in King and Gilbert, *The Letters of Teilhard de Chardin and Lucile Swan*, p. xv.

155 **comparison of *Sinanthropus* with modern humans:** Drawn from the description in Shapiro, *Peking Man*, pp. 58–73.

155 **taller, on average, than Peking Man:** Shapiro, *Peking Man*, p. 58.

157 **dragged to their cave to eat:** Boaz and Ciochon, *Dragon Bone Hill*, p. 130.

157 **Weidenreich's contention:** Tattersall, *The Fossil Trail*, p. 62.

158 **retire to comfortably at night:** Arnould, *Teilhard de Chardin*, p. 185. Author's translation.

158 **". . . chance of an individual variation":** TdC to MB, August 2, 1930, Teilhard de Chardin, *Teilhard de Chardin en Chine*, p. 212. Author's translation.

159 **found both in Africa and in Europe:** See, for example, Schwartz and Tattersall, "What Constitutes *Homo erectus?*" pp. 21–25.

CHAPTER 12 • THE YELLOW CRUISE AND THE MONGOLIAN PRINCESS

161 **Croisière Jaune, the Yellow Cruise:** Some of the details in this chapter come from the Citroën website: http://www.citroen.com/CWW/en-US/HISTORY/ADVENTURE/YellowCruise/.

162 **Marco Polo almost seven centuries earlier:** Commentary by Claude Aragonnès [Marguerite Teillard-Chambon], in Teilhard de Chardin, *Lettres de Voyage (1923–1955)*, p. 190.

162 **"golden altars, without idols":** Arnould, *Teilhard de Chardin*, p. 187. Author's translation.

163 **". . . is extremely 'captivating' ":** TdC to MTC, May 22, 1931, Teilhard de Chardin, *Lettres de Voyage (1923–1955)*, pp. 194–196. Author's translation.

164 **". . . the spirit of the expedition":** Arnould, *Teilhard de Chardin*, p. 189. Author's translation. Haardt died a month after the expedition ended.

164 **But the team pressed on:** Le Fèvre, *La Croisière Jaune*, pp. 112–113. Exchanges of dialogue in this description are from Le Fèvre's account; author's translation.

165 **embroiled in a bitter civil war:** Commentary by Claude Aragonnès [Marguerite Teillard-Chambon], in Teilhard de Chardin, *Lettres de Voyage (1923–1955)*, p. 150.

165 **". . . deepest and most virginal state":** TdC to MTC, July 7, 1931, ibid., pp. 151–152. Author's translation.

165 **"one of the most sacred and mysterious" geological regions of Asia:** Lukas and Lukas, *Teilhard*, p. 124.

166 **". . . country that is so poor . . . difficult roads":** Le Fèvre, *La Croisière Jaune*, p. 241.

166 **". . . adventures in Sinkiang":** TdC to MTC, August 27, 1931, Teilhard de Chardin, *Lettres de Voyage (1923–1955)*, pp. 198–199. Author's translation.

167 **a statue of Nirgidma:** Arnould, *Teilhard de Chardin*, p. 190. A photograph of the portrait sculpture of Nirgidma by Lucile Swan can be found in King and Gilbert, *The Letters of Teilhard de Chardin and Lucile Swan*, p. 144.

167 **"Parlez-moi d'amour . . .":** Arnould, *Teilhard de Chardin*, p. 190.

168 **". . . like a rock of marble":** Quoted in Speaight, *The Life of Teilhard de Chardin*, p. 184.

168 **Teilhard hypothesized:** Speaight, *The Life of Teilhard de Chardin*, p. 186.

168 **geological unity of central Asia:** Le Fèvre, *La Croisière Jaune*, p. 271.

170 **". . . this new year . . .":** Quoted in Arnould, *Teilhard de Chardin*, p. 191 (from Louis Auduin-Dubreuil, *Sur la Route de la Soie* [Paris: Plon, 1935], p. 225). Author's translation.

170 **"... attention to our expedition"**: TdC to MTC, January 30, 1932, Teilhard de Chardin, *Lettres de Voyage (1923–1955)*, pp. 206–208. Author's translation.

171 **"General of Independent Cavalry"**: Commentary by Claude Aragonnès [Marguerite Teillard-Chambon], based on information in Le Fèvre, *La Croisière Jaune*, pp. 329–330, in Teilhard de Chardin, *Lettres de Voyage (1923–1955)*, pp. 206–208. Author's translation.

171 **"... a price to pay for this"**: Letter dated February 12, 1932, Arnould, *Teilhard de Chardin*, p. 193. Author's translation.

171 **"... upon which everything is built"**: Ibid. Author's translation.

172 **Congress to be held in July:** Arnould, *Teilhard de Chardin*, p. 194.

173 **have corroborated Teilhard's analysis:** See Weiner et al., "Evidence for the Use of Fire at Zhoukoudian," and Goldberg et al., "Site Formation Processes at Zhoukoudian, China."

174 **"Our colleague ... for two?" ... was sleeping alone:** Ibid., p. 198. Author's translation.

174 **"... at home in California"**: TdC to MTC, September 12, 1933, Arnould, *Teilhard de Chardin*, p. 198. Author's translation.

CHAPTER 13 · LUCILE SWAN RECONSTRUCTS PEKING MAN

175 **"... 'cavern' of the *Sinanthropus*"**: TdC to HB, December 31, 1933, Teilhard de Chardin, *Lettres à l'Abbé Gaudefroy et l'Abbé Breuil*, pp. 197–198. Author's translation.

176 **"... known through its eggs"**: TdC to MB, February 15, 1934, Teilhard de Chardin, *Teilhard de Chardin en Chine*, p. 234. The report that he mentions is by Pei Wenzhong and others: "A Preliminary Report on the Late Paleolithic Cave of Choukoutien."

178 **"... will become more apparent"**: TdC to LS, March 18, 1934, King and Gilbert, *The Letters of Teilhard de Chardin and Lucile Swan*, pp. 10–11.

178 **what had to be done now:** Barbour, *In the Field with Teilhard de Chardin*, pp. 66–67.

179 **carefully written descriptions:** Reported in Walker and Shipman, *The Wisdom of the Bones*, p. 67.

182 **"... prematurely unfrocked?"**: Speaight, *The Life of Teilhard de Chardin*, p. 202.

183 "... few patches only of the deposits are preserved": TdC to LS, January 28, 1935, King and Gilbert, *The Letters of Teilhard de Chardin and Lucile Swan*, p. 26.

183 He ended that letter ... with the following: TdC to LS, February 8, 1935, ibid., p. 27.

184 "... worthwhile line of effort—": LS, journal entry, February 14, 1935, ibid., p. 28.

184 "... God bless you, precious / Pierre": TdC to LS, March 29, 1935, ibid., p. 29.

185 "... each blossom of tree, I think": TdC to LS, April 6–7, 1935, ibid., pp. 29–30.

185 continuously for more than fifteen days: Madeleine Barthélemy-Madaule, introduction to Teilhard de Chardin, *Lettres de Voyage (1923–1955)*, p. 7.

186 "... see her at the end of the month": TdC to LS, June 16, 1935, King and Gilbert, *The Letters of Teilhard de Chardin and Lucile Swan*, pp. 38–39.

188 diverged from that of chimpanzees: Around 2 million years ago, chimpanzees diverged into the common chimpanzee and the bonobo.

CHAPTER 14 · PEKING MAN VANISHES

191 accompanied him to his lectures: Speaight, *The Life of Teilhard de Chardin*, p. 230.

191 common ancestry of humans and apes: William L. Laurence, "China Cave a Lead to 'Missing Link,'" *The New York Times*, March 20, 1937, p. 9.

193 "... whistle in their tails": TdC to Pierre Lamare, April 25, 1941, Speaight, *The Life of Teilhard de Chardin*, p. 180.

195 "... writing a book on [Abraham] Lincoln": TdC to LS, May 29, 1939, King and Gilbert, *The Letters of Teilhard de Chardin and Lucile Swan*, pp. 136–137.

196 "... what's it all about anyway": LS to TdC, October 1939, ibid., p. 139.

197 Imperial Household Museum in Tokyo: Shapiro, *Peking Man*, pp. 16–17.

198 Father Edmund Walsh at Georgetown University: Lukas and Lukas, *Teilhard*, p. 181.

199 "... but I miss you": TdC to LS, August 11, 1941, King and Gilbert, *The Letters of Teilhard de Chardin and Lucile Swan*, p. 143.

199 **". . . we will be together again":** Ibid., p. 153.

200 **Time was running out:** Shapiro, *Peking Man*, pp. 18–19.

200 **Japanese were in control of the city:** Ibid., pp. 19–20.

200 **took over Dragon Bone Hill back in 1937:** Boaz and Ciochon, *Dragon Bone Hill*, p. 47.

200 **five days of continuous interrogation:** Shapiro, *Peking Man*, p. 15.

201 **But they, too, were unsuccessful:** Ibid., p. 20.

201 **more resources to the search for the remains:** Boaz and Ciochon, *Dragon Bone Hill*, p. 47.

202 **on the grounds of the college or outside it:** Ibid., p. 46.

<h3 style="text-align:center">CHAPTER 15 · ROME</h3>

203 **". . . And I DO DO DO miss you!":** TdC to LS, November 26, 1941, King and Gilbert, *The Letters of Teilhard de Chardin and Lucile Swan*, p. 154.

204 **which he did . . . on December 28:** Lukas and Lukas, *Teilhard*, p. 204.

205 **". . . Yours, as before! +++ / P.T.":** TdC to LS, August 31, 1945, King and Gilbert, *The Letters of Teilhard de Chardin and Lucile Swan*, p. 163.

205 **"With much love, P.T.":** TdC to LS, December 20, 1945, ibid., p. 174.

205 **". . . All my love dearest":** LS to TdC, January 4, 1946, ibid., p. 175.

207 **". . . piling up on my table. . . .":** TdC to LS, June 18, 1946, ibid., pp. 185–186.

207 **hear his conversations:** Speaight, *The Life of Teilhard de Chardin*, p. 275.

208 **objections raised by the censors in Rome:** Ibid., p. 271.

209 **". . . submission to the Church":** Ibid.

210 **"many floating mines":** TdC to Léontine Zanta, June 24, 1934, Teilhard de Chardin, *Letters to Léontine Zanta*, p. 110.

210 **". . . a transcendent God":** Speaight, *The Life of Teilhard de Chardin*, pp. 278–279.

211 **". . . each one has a car":** Thomas M. King, S.J., personal communication to author, June 6, 2006.

211 **help in getting around:** Thomas M. King, S.J., personal communication to author, December 20, 2006.

211 **". . . in Washington till Wednesday":** TdC to LS, King and Gilbert, *The Letters of Teilhard de Chardin and Lucile Swan*, p. 226. This letter, unlike others, was written in French; the translation is King and Gilbert's.

212 **". . . I built my life upon it":** Ibid., p. 227.

213 **"... to be truly pleasant":** TdC to Jeanne Mortier, October 8, 1948, Teilhard de Chardin, *Lettres à Jeanne Mortier*, pp. 43–44. Author's translation.

214 **"... God bless you, dearest!":** TdC to LS, September 3, 1948, King and Gilbert, *The Letters of Teilhard de Chardin and Lucile Swan*, pp. 235–236.

214 **"... Have a good time in Berne! . . . Teilhard":** TdC to LS, October 7, 1948, ibid., pp. 236–237.

215 **"... after 3 p.m.":** TdC to LS, October 1948, ibid., p. 237.

216 **the pagan symbol it was:** Speaight, *The Life of Teilhard de Chardin*, p. 283.

216 **speak to him about the Auvergne:** Ibid., p. 284.

216 **"... there is no problem":** Ibid., p. 284.

217 **"Man's Sense of and Place in Nature"; "Man and the Definition of Species":** Arnould, *Teilhard de Chardin*, pp. 320–321.

218 **ordered . . . to leave their positions:** Lukas and Lukas, *Teilhard*, p. 285.

CHAPTER 16 • AFTERMATH

220 **live under strict surveillance:** Lukas and Lukas, *Teilhard*, p. 303.

221 **"... so I mothered him":** Thomas M. King, S.J., personal communication to author, June 6, 2006.

223 **"... than the preceding one . . .":** TdC to Jeanne Mortier, August 14, 1953, Teilhard de Chardin, *Lettres à Jeanne Mortier*, pp. 78–79. Author's translation.

223 **"... without any provocation":** TdC to Jeanne Mortier, September 10, 1953, ibid., p. 81. Author's translation.

224 **involved in this fraud:** Stephen Jay Gould wrote several articles about his suspicion that Teilhard was involved in the Piltdown Man hoax; two notable ones appeared in the March 1979 and June 1981 issues of *Natural History*.

224 **"... totality of our experience":** Crespy, *From Science to Theology*, pp. 30–31.

224 **disturbing to the Jesuit leaders:** Speaight, *The Life of Teilhard de Chardin*, p. 323.

225 **"... Yours, very affectueusement, Pierre":** TdC to LS, March 30, 1955, King and Wood, *The Letters of Teilhard de Chardin and Lucile Swan*, pp. 293–294.

226 **none of his friends . . . was present:** Ibid., p. 332.

226 **to great acclaim:** The titles in French are: *Le Phénomène Humain, Lettres de Voyage, Le Milieu Divin,* and *L'Avenir de l'Homme.*

227 **"'. . . And I shall leave it at that' ":** Dobzhansky, *Mankind Evolving,* p. 348.

CHAPTER 17 · THE FOSSIL RECORD CONTINUES

231 **". . . for another million years":** Johanson, Johanson, and Edgar, *Ancestors,* p. 42.

232 **Middle Awash Valley . . . 4.1 million years ago:** White et al., "Assa Issie, Aramis and the Origin of *Australopithecus.*"

232 **news from Dikika . . . fossils of a girl:** Alemseged et al., "A Juvenile Early Hominin Skeleton from Dikika, Ethiopia."

233 **the outskirts of Europe:** Rightmire et al., "Anatomical Descriptions, Comparative Studies and Evolutionary Significance of the Hominid Skulls from Dmanisi, Republic of Georgia."

233 **human lineage from that of chimpanzees:** Zollikofer et al., "Virtual Cranial Reconstruction of *Sahelanthropus tchadensis,*" p. 755.

235 **30,000 to 50,000 years ago:** Swisher et al., "Latest *Homo erectus* of Java."

236 **must imply the use of language:** Richard E. Leakey, *The Making of Mankind,* p. 137.

237 **roots of modern languages:** Ibid., p. 139.

238 **shells with holes . . . 90,000 years ago:** Vanhaeren et al., "Middle Paleolithic Shell Beads in Israel and Algeria."

CHAPTER 18 · WHAT REALLY HAPPENED TO PEKING MAN?

240 **transported them to Japan:** Shapiro, *Peking Man,* p. 22.

240 **original Peking Man finds:** Ibid., pp. 24–25.

241 **letter denying the claim:** Ibid., p. 26.

245 **Skull XI in the Weidenreich list:** Ibid., p. 167.

245 **Seng . . . the woman's attorney:** Janus with Brashler, *The Search for Peking Man,* p. 111.

246 **it was probably false:** Ibid., p. 205.

246 **dumped . . . into the Yellow Sea:** Ibid., p. 208.

246 **$640,000 . . . to personal use:** Jia and Huang, *The Story of Peking Man,* p. 182.

247 **colleagues found nothing:** Ibid., p. 182.

248 **"... ship was not sunk":** Ibid., pp. 183–184.

249 **"... some courtyard in Peking":** Arnould, *Teilhard de Chardin*, p. 184. Author's translation.

251 **"'... We are going forward!' ":** Teilhard de Chardin, *The Future of Man*, p. 1.

251 **"... others go beyond me":** As quoted in Cuénot, *Science and Faith in Teilhard de Chardin*, p. vii.

APPENDIX 2 · RADIOCARBON AND OTHER
SCIENTIFIC DATING METHODS

257 **stopped absorbing carbon 14:** See Bowman, *Radiocarbon Dating*, pp. 9–24.

REFERENCES

Aczel, Amir D. *God's Equation: Einstein, Relativity, and the Expanding Universe*. New York: Four Walls Eight Windows, 1999.

Aczel, Amir D. "Improved Radiocarbon Age Estimation Using the Bootstrap." *Radiocarbon* 37 (1995), pp. 845–849.

Aczel, Amir D. *Pendulum: Léon Foucault and the Triumph of Science*. New York: Atria Books, 2003.

Aiello, Leslie, and Christopher Dean. *An Introduction to Human Evolutionary Anatomy*. San Diego: Academic Press, 1990.

Alemseged, Zeresenay, et al. "A Juvenile Early Hominin Skeleton from Dikika, Ethiopia." *Nature* 443 (September 21, 2006), pp. 296–301.

Andersson, J. Gunnar. *Children of the Yellow Earth: Studies in Prehistoric China*. Translated by E. Classen. Cambridge, MA: MIT Press, 1973.

Arensburg, B., et al. "A Middle Paleolithic Human Hyoid Bone." *Nature* 338 (1989), pp. 758–760.

Arnould, Jacques. *Teilhard de Chardin*. Paris: Perrin, 2005.

Arsuaga, Juan Luis. *The Neanderthal's Necklace: In Search of the First Thinkers*. New York: Four Walls Eight Windows, 2002.

Auduin-Dubreuil, Louis. *Sur la Route de la Soie*. Paris: Plon, 1935.

Bar-Yosef, Ofer. "On the Nature of Transitions: The Middle to Upper Paleolithic and the Neolithic Revolution." *Cambridge Archaeological Journal* 8 (1998), pp. 141–163.

Bar-Yosef, Ofer. "The Role of Western Asia in Modern Human Origins." *Philosophical Transactions of the Royal Society* B 337 (1992), pp. 193–200.

Bar-Yosef, Ofer, and Jane Callander. "Dorothy Annie Elizabeth Garrod (1892–1968)." In Getzel M. Cohen and Martha Sharp Joukowsky, eds., *Breaking Ground: Pioneering Women Archaeologists.* Ann Arbor: University of Michigan Press, 2004.

Barbour, George. *In the Field with Teilhard de Chardin.* New York: Herder and Herder, 1965.

Behe, Michael J. *Darwin's Black Box: The Biochemical Challenge to Evolution.* New York: Free Press, 1996.

Bergson, Henri. *Creative Evolution.* Translated by Arthur Mitchell. New York: Henry Holt, 1911.

Berra, Tim M. *Evolution and the Myth of Creationism.* Stanford, CA: Stanford University Press, 1990.

Binford, L. R. *Bones: Ancient Men and Modern Myths.* New York: Academic Press, 1981.

Boaz, Noel T., and Russell L. Ciochon. *Dragon Bone Hill: An Ice-Age Saga of Homo erectus.* New York: Oxford University Press, 2004.

Boule, Marcellin, and Henri Vallois. *Fossil Men: A Textbook of Human Palaeontology.* London: Thames & Hudson, 1957.

Bowler, Peter J. *Theories of Human Evolution: A Century of Debate, 1844–1944.* Baltimore: Johns Hopkins University Press, 1986.

Bowman, Sheridan. *Radiocarbon Dating: Interpreting the Past.* Berkeley: University of California Press, 1990.

Brain, C. K. *The Hunters or the Hunted? An Introduction to African Cave Taphonomy.* Chicago: University of Chicago Press, 1981.

Bramble, Dennis M., and Daniel E. Lieberman. "Endurance Running and the Evolution of *Homo.*" *Nature* 432 (November 18, 2004), pp. 345–351.

Brunet, Michel, et al. "New Material of the Earliest Hominid from the Upper Miocene of Chad." *Nature* 434 (April 7, 2005), pp. 752–755.

Callander, Jane, and Ofer Bar-Yosef. "Saving Mount Carmel Caves: A Cautionary Tale for Archaeology in Our Times." *Palestine Exploration Quarterly* 132 (2000), pp. 94–111.

Chang, Kwang-chih. *The Archaeology of Ancient China*. New Haven, CT: Yale University Press, 1968.

Cherry, Michael. "Genetics to Unlock Secrets of Our African Past," *Nature* 422 (April 3, 2003), p. 460.

Conroy, Glenn C. *Primate Evolution*. New York: W. W. Norton, 1990.

Corte, Nicolas. *Pierre Teilhard de Chardin: His Life and Spirit*. Translated by Martin Jarrett-Kerr. New York: Macmillan, 1960.

Crespy, Georges. *From Science to Theology: An Essay on Teilhard de Chardin*. Translated by George H. Shriver. New York: Abingdon Press, 1968.

Cuénot, Claude. *Ce Que Teilhard A Vraiment Dit*. Paris: Stock, 1972.

Cuénot, Claude. *Science and Faith in Teilhard de Chardin*. Tiptree, England: Anchor, 1967.

Cuénot, Claude. *Teilhard de Chardin*. Paris: Plon, 1958.

Cuénot, Claude. *Teilhard de Chardin: A Biographical Study*. Translated by Vincent Colimore. London: Burns & Oates, 1965.

Currie, Pete. "Muscling In on Hominid Evolution." *Nature* 428 (March 25, 2004), pp. 373–374.

Dalton, Rex. "Anthropologists Walk Tall After Unearthing Hominid." *Nature* 434 (March 10, 2005), p. 126.

Dart, Raymond A., with Dennis Craig. *Adventures with the Missing Link*. Philadelphia: Institutes Press, 1959.

Darwin, Charles. *The Descent of Man* (*The Descent of Man, and Selection in Relation to Sex*, 1871). London: Gibson Square, 2003.

Darwin, Charles. *The Origin of Species* (*On the Origin of Species by Means of Natural Selection*, 1859). New York: Barnes & Noble, 2004.

Darwin, Charles. *The Voyage of the Beagle* (1839). New York: Dover, 2002.

Dawkins, Richard. *The Ancestor's Tale*. New York: Houghton Mifflin, 2004.

Dawkins, Richard. *River Out of Eden*. London: Weidenfeld & Nicolson, 1995.

Dawkins, Richard. *The Selfish Gene*. 2nd ed. Oxford: Oxford University Press, 1989.

Day, Michael H. *Guide to Fossil Man*. 4th edition. Chicago: University of Chicago Press, 1986.

de la Héronnière, Edith. *Teilhard de Chardin: Une Mystique de la Traverse*. Paris: Albin Michel, 2003.

de Lubac, Henri. *La Pensée Religieuse du Père Pierre Teilhard de Chardin*. Paris: Aubier, 1962.

de Terra, Helmut. *Mes Voyages avec Teilhard de Chardin*. Paris: Seuil, 1965.

Diamond, Jared. *The Third Chimpanzee: The Evolution and Future of the Human Animal*. New York: HarperCollins, 1992.

Dobzhansky, Theodosius. *Evolution, Genetics, and Man*. New York: John Wiley & Sons, 1955.

Dobzhansky, Theodosius. *Genetics and the Origin of Species*. New York: Columbia University Press, 1937.

Dobzhansky, Theodosius. *Mankind Evolving: The Evolution of the Human Species*. New Haven, CT: Yale University Press, 1962.

d'Ouince, René. *Un Prophète en Procès: Teilhard de Chardin et l'Église de Son Temps*. Paris: Aubier-Montaigne, 1970.

Dumiak, Michael. "The Neanderthal Code." *Archaeology*, November/December 2006, pp. 22–25.

Eldredge, Niles. *Time Frames: The Rethinking of Darwinian Evolution and the Theory of Punctuated Equilibria*. New York: Simon & Schuster, 1985.

Eldredge, Niles, and Joel Cracraft. *Philogenetic Patterns and the Evolutionary Process*. New York: Columbia University Press, 1980.

Finlayson, Clive. *Neanderthals and Modern Humans: An Ecological and Evolutionary Perspective*. New York: Cambridge University Press, 2004.

Fisher, Ronald A. *The Genetical Theory of Natural Selection*. Oxford, England: Oxford University Press, 1930.

Fleagle, John G. *Primate Adaptation & Evolution*. San Diego: Academic Press, 1988.

Foley, Robert. *Another Unique Species*. Harlow, England: Longman, 1987.

Frere, John. "An Account of Flint Weapons Discovered at Hoxne in Suffolk." *Archaeologia* 13 (1800), pp. 204–205.

Gibbons, Ann. *The First Human*. New York: Doubleday, 2006.

Goldberg, Paul, et al. "Site Formation Processes at Zhoukoudian, China." *Journal of Human Evolution* 41 (2001), pp. 483–530.

Gould, Stephen Jay. *Ever Since Darwin: Reflections in Natural History*. New York: W.W. Norton, 1977.

Gould, Stephen Jay. *The Structure of Evolutionary Theory*. Cambridge, MA: Harvard University Press, 2002.

Grandclément, Daniel. *L'Incroyable Henry de Monfreid*. Paris: Grasset, 1998.

Gribbin, John, and Jeremy Cherfas. *The First Chimpanzee: In Search of Human Origins*. London: Penguin, 2001.

Grine, Frederick E., ed. *Evolutionary History of the "Robust" Australopithecines*. New York: Aldine de Gruyter, 1988.

Grün, Rainer, et al. "ESR Analysis of Teeth from the Palaeoanthropological Site of Zhoukoudian, China." *Journal of Human Evolution* 32 (1997), pp. 83–91.

Hood, Dora. *Davidson Black: A Biography*. Toronto: University of Toronto Press, 1964.

Huxley, Leonard. *The Life and Letters of Thomas Henry Huxley*. New York: D. Appleton, 1901.

Huxley, Thomas Henry. *Man's Place in Nature* (1863). Ann Arbor: University of Michigan Press, 1959.

Isaac, Glynn L., and Elizabeth R. McCown, eds. *Human Origins: Louis Leakey and the East African Evidence*. Menlo Park, CA: W. A. Benjamin, 1976.

Janus, Christopher G., with William Brashler. *The Search for Peking Man*. New York: Macmillan, 1975.

Jepsen, Glenn L., Ernst Mayr, and George Gaylord Simpson, eds. *Genetics, Paleontology, and Evolution*. Princeton, NJ: Princeton University Press, 1949.

Jerison, Harry J. *Evolution of the Brain and Intelligence*. New York: Academic Press, 1973.

Jia, Lanpo. *Early Man in China*. Beijing: Foreign Languages Press, 1980.

Jia, Lanpo, and Weiwen Huang. *The Story of Peking Man: From Archaeology to Mystery*. Translated by Yin Zhinqui. New York: Oxford University Press, 1990.

Johanson, Donald C., and Maitland A. Edey. *Lucy: The Beginnings of Humankind*. New York: Simon & Schuster, 1981.

Johanson, Donald C., and Blake Edgar. *From Lucy to Language*. New York: Simon & Schuster, 1996.

Johanson, Donald, Lenora Johanson, and Blake Edgar. *Ancestors: In Search of Human Origins*. New York: Villard, 1994.

Johanson, Donald, and James Shreeve. *Lucy's Child: The Discovery of a Human Ancestor*. New York: William Morrow, 1982.

Joint Expedition of the British School of Archaeology in Jerusalem and the American School of Prehistoric Research (1929–1934). *The Stone Age of Mount Carmel*. Oxford: Clarendon Press, 1937.

Jones, Steve, Robert Martin, and David Pilbeam, eds. *The Cambridge Encyclopedia of Human Evolution*. Cambridge, England: Cambridge University Press, 1992.

Kennedy, G. E. *Paleoanthropology*. New York: McGraw-Hill, 1980.

King, Thomas M. *Teilhard's Mass: Approaches to "The Mass on the World."* New York: Paulist Press, 2005.

King, Thomas M. *Teilhard's Mysticism of Knowing*. New York: Seabury, 1981.

King, Thomas M., and Mary Wood Gilbert, eds., *The Letters of Teilhard de Chardin and Lucile Swan*. Washington, DC: Georgetown University Press, 1993.

King, Thomas M., and James F. Salomon, eds. *Teilhard and the Unity of Knowledge*. New York: Paulist Press, 1983.

King, Ursula. *Spirit of Fire: The Life and Vision of Teilhard de Chardin*. Maryknoll, NY: Orbis Books, 1996.

Klein, Jan, and Naoyuki Takahata. *Where Do We Come From: The Molecular Evidence for Human Descent*. New York: Springer, 2002.

Klein, Richard G. *The Human Career: Human Biological and Cultural Origins*. Chicago: University of Chicago Press, 1989.

Klein, Richard, with Blake Edgar. *The Dawn of Human Culture*. New York: John Wiley & Sons, 2002.

Koenigswald, G. H. R. von. *Meeting Prehistoric Man*. London: Thames & Hudson, 1956.

Landau, Misia. *Narratives of Human Evolution*. New Haven, CT: Yale University Press, 1991.

Lane, David H. *The Phenomenon of Teilhard: Prophet for a New Age*. Macon, GA: Mercer University Press, 1996.

Le Fèvre, Georges. *La Croisière Jaune: Expédition Citroën Centre-Asie*. Paris: Plon, 1933.

Leakey, Louis S. B. *By the Evidence: Memoirs, 1932–1951*. New York: Harcourt Brace Jovanovich, 1974.

Leakey, Louis S. B. *White African*. London: Hodder & Stoughton, 1937.

Leakey, M. D. *Olduvai Gorge*, vol. 3. Cambridge, England: Cambridge University Press, 1971.

Leakey, M. D., and J. M. Harris, eds. *Laetoli: A Pliocene Site in Northern Tanzania*. Oxford, England: Clarendon Press, 1987.

Leakey, Richard E. *The Making of Mankind*. London: Sphere Books, 1982.

Leakey, Richard E. *The Origin of Humankind*. New York: Basic Books, 1994.

Leakey, Richard E., and Roger Lewin. *Origins Reconsidered*. New York: Doubleday, 1992.

Leakey, Richard E., and Roger Lewin. *People of the Lake: Mankind and Its Beginnings*. New York: Doubleday, 1978.

Lenski, G. *Ecological-Evolutionary Theory: Principles and Applications*. Boulder, CO: Paradigm, 2005.

Leroy, Pierre. *Lettres Familières de Pierre Teilhard de Chardin Mon Ami: Les Dernières Années (1948–1955)*. Paris: Le Centurion, 1976.

Levy, Leonard W., ed. *The World's Most Famous Trial: State of Tennessee v. John Thomas Scopes*. New York: Da Capo, 1971.

Lewin, Roger. *Bones of Contention: Controversies in the Search for Human Origins*. New York: Simon & Schuster, 1987.

Lieberman, Philip. *The Biology and Evolution of Language*. Cambridge, MA: Harvard University Press, 1984.

Lukas, Mary, and Ellen Lukas. *Teilhard*. New York: Doubleday, 1977.

Mayr, Ernst. *The Growth of Biological Thought: Diversity, Evolution, and Inheritance*. Cambridge, MA: Belknap Press, 1982.

Mayr, Ernst. *Systematics and the Origin of Species, from the Viewpoint of a Zoologist*. New York: Columbia University Press, 1942.

McCown, Theodore D., and Kenneth A. R. Kennedy, eds. *Climbing Man's Family*

Tree: A Collection of Major Writings on Human Phylogeny, 1699 to 1971. Englewood Cliffs, NJ: Prentice-Hall, 1972.

McKee, Jeffrey K. *The Riddled Chain: Chance, Coincidence, and Chaos in Human Evolution.* New Brunswick, NJ: Rutgers University Press, 2000.

Megarry, Tim. *Society in Prehistory: The Origins of Human Culture.* New York: New York University Press, 1995.

Mellars, Paul. *The Emergence of Modern Humans.* Ithaca, NY: Cornell University Press, 1990.

Mellars, Paul. *The Neanderthal Legacy: An Archaeological Perspective of Western Europe.* Princeton, NJ: Princeton University Press, 1996.

Mellars, Paul. "A New Radiocarbon Revolution and the Dispersal of Modern Humans in Eurasia." *Nature* 439 (February 23, 2006), pp. 931–935.

Mellars, Paul, and Kathleen Gibson, eds. *Modelling the Early Human Mind.* Exeter, England: Short Run Press, 1996.

Mithen, Steven. *After the Ice: A Global Human History 20,000–5,000 B.C.* London: Weidenfeld & Nicolson, 2003.

Nemeck, Francis Kelly. *Teilhard de Chardin et Jean de la Croix.* Montreal: Bellarmin, 1975.

Nitecki, Matthew H., and Doris V. Nitecki, eds. *Origins of Anatomically Modern Humans.* New York: Plenum Press, 1994.

O'Connor, Catherine R. *Woman and Cosmos: The Feminine in the Thought of Pierre Teilhard de Chardin.* Englewood Cliffs, NJ: Prentice-Hall, 1971.

Otte, Daniel, and John A. Endler, eds. *Speciation and Its Consequences.* Sunderland, MA: Sinauer, 1989.

Pei, Wenzhong, et al. "A Preliminary Report on the Late Paleolithic Cave of Choukoutien." *Bulletin of the Geological Society of China* 13 (1934), pp. 327–358.

Pfeiffer, John E. *The Emergence of Man.* New York: Harper & Row, 1972.

Pilbeam, David. *The Ascent of Man: An Introduction to Human Evolution.* New York: Macmillan, 1972.

Pinker, Steven. *How the Mind Works.* New York: W. W. Norton, 1997.

Pinker, Steven. *The Language Instinct: How the Mind Creates Language*. New York: William Morrow, 1994.

Piveteau, Jean. *Le Père Teilhard de Chardin, Savant*. Paris: Fayard, 1964.

Portmann, Adolf. *A Zoologist Looks at Humankind*. Translated by Judith Schaefer. New York: Columbia University Press, 1990.

Reader, John. *Man on Earth*. London: Collins, 1988.

Reader, John. *Missing Links: The Hunt for Earliest Man*. Boston: Little, Brown, 1981.

Regal, Brian. *Human Evolution: A Guide to the Debates*. Santa Barbara, CA: ABC-CLIO, 2004.

Rideau, Émile. *La Pensée du Père Teilhard de Chardin*. Paris: Seuil, 1965.

Rightmire, G. Philip, et al. "Anatomical Descriptions, Comparative Studies and Evolutionary Significance of the Hominid Skulls from Dmanisi, Republic of Georgia." *Journal of Human Evolution* 50 (2006), pp. 115–141.

Rivière, Claude. *En Chine avec Teilhard (1938–1944)*. Paris: Seuil, 1968.

Roberts, Noel K. *From Piltdown Man to Point Omega: The Evolutionary Theory of Teilhard de Chardin*. New York: Lang, 2000.

Rohde, Douglas L. T., Steve Olson, and Joseph T. Chang. "Modelling the Recent Common Ancestry of All Living Humans." *Nature* 431 (September 30, 2004), pp. 562–569.

Schick, Kathy D., and Nicholas Toth. *Making Silent Stones Speak: Human Evolution and the Dawn of Technology*. New York: Simon & Schuster, 1993.

Schwartz, Jeffrey H. *The Red Ape: Orangutans and Human Origins*. Cambridge, MA: Westview, 2005.

Schwartz, Jeffrey H., and Ian Tattersall. "What Constitutes *Homo erectus*?" *Acta Anthropologica Sinica*, suppl. to vol. 19 (2000), pp. 21–25.

Scopes, John T., and James Presley. *Center of the Storm: Memoirs of John T. Scopes*. New York: Holt, Rinehart and Winston, 1967.

Shapiro, Harry L. *Peking Man*. New York: Simon & Schuster, 1974.

Shipman, Pat. *The Man Who Found the Missing Link: Eugène Dubois and His Lifelong Quest to Prove Darwin Right*. New York: Simon & Schuster, 2001.

Shreeve, James. *The Neandertal Enigma: Solving the Mystery of Modern Human Origins.* New York: William Morrow, 1995.

Sigmon, Becky A., and Jerome S. Cybulski, eds. *Homo erectus: Papers in Honor of Davidson Black.* Toronto: University of Toronto Press, 1981.

Smith, Fred H., and Frank Spencer, eds. *The Origins of Modern Humans: A World Survey of the Fossil Evidence.* New York: Alan R. Liss, 1984.

Speaight, Robert. *The Life of Teilhard de Chardin.* New York: Harper & Row, 1967.

Stringer, Christopher, and Clive Gamble. *In Search of the Neanderthals: Solving the Puzzle of Human Origins.* New York: Thames & Hudson, 1993.

Stringer, Christopher, and Robin McKie. *African Exodus: The Origins of Modern Humanity.* New York: Henry Holt, 1997.

Swisher, Carl C., III, Garniss H. Curtis, and Roger Lewin. *Java Man.* New York: Scribner, 2000.

Swisher, Carl C., III, et al., "Latest *Homo erectus* of Java." *Science* 274 (1996), pp. 1870–1874.

Tattersall, Ian. *Becoming Human: Evolution and Human Uniqueness.* New York: Harcourt, 1998.

Tattersall, Ian. *The Fossil Trail: How We Know What We Think We Know About Human Evolution.* New York: Oxford University Press, 1995.

Tattersall, Ian. *The Human Odyssey: Four Million Years of Human Evolution.* New York: Prentice-Hall, 1993.

Tattersall, Ian. *The Monkey in the Mirror: Essays on the Science of What Makes Us Human.* New York: Harcourt, 2002.

Tattersall, Ian, E. Delson, and J. A. Van Couvering, eds. *Encyclopedia of Human Evolution and Prehistory.* New York: Garland, 1988.

Tattersall, Ian, and G. Sawyer. "The Skull of 'Sinanthropus' from Zhoukoudian, China: A New Reconstruction." *Journal of Human Evolution* 31 (1996), pp. 311–314.

Teilhard de Chardin, Pierre. *Early Man in China.* Peking (Beijing): Institute of Geobiology, 1941.

Teilhard de Chardin, Pierre. *The Future of Man.* Translated by Norman Denny. New York: Doubleday, 2004.

Teilhard de Chardin, Pierre. *Genèse d'une Pensée: Lettres (1914–1919)*. Paris: Grasset, 1961.

Teilhard de Chardin, Pierre. *Images et Paroles*. Paris: Seuil, 1966.

Teilhard de Chardin, Pierre. *Journal*, vol. 1 (August 26, 1915–January 4, 1919). Paris: Fayard, 1975.

Teilhard de Chardin, Pierre. *Letters to Léontine Zanta*. Translated by Bernard Wall. New York: Harper & Row, 1968.

Teilhard de Chardin, Pierre. *Letters from a Traveller*. Edited and translated (from *Lettres de Voyage [1923–1955]*) by René Hague et al., with introductions by Julian Huxley and Pierre Leroy. London: Collins, 1962.

Teilhard de Chardin, Pierre. *Letters to Two Friends, 1926–1952*. New York: New American Library, 1967.

Teilhard de Chardin, Pierre. *Lettres à l'Abbé Gaudefroy et l'Abbé Breuil*. Paris: Rocher, 1988.

Teilhard de Chardin, Pierre. *Lettres d'Égypte (1905–1908)*. Paris: Aubier-Montaigne, 1963.

Teilhard de Chardin, Pierre. *Lettres Intimes à Auguste Valensin, Bruno de Solages, Henri de Lubac, André Ravier*. Paris: Aubier-Montaigne, 1974.

Teilhard de Chardin, Pierre. *Lettres à Jeanne Mortier*. Paris: Seuil, 1984.

Teilhard de Chardin, Pierre. *Lettres de Voyage (1923–1955)*. Edited by Claude Aragonnès. Paris: Maspero, 1982.

Teilhard de Chardin, Pierre. *Notes de Retraites (1919–1954)*. Paris: Seuil, 2003.

Teilhard de Chardin, Pierre. *Nouvelles Lettres de Voyage (1939–1955)*. Paris: Grasset, 1957.

Teilhard de Chardin, Pierre. *Oeuvres*, 13 vols. Paris: Seuil, 1955–1976.

Teilhard de Chardin, Pierre. *Le Phénomène Humain*. Paris: Seuil, 1955.

Teilhard de Chardin, Pierre. *Teilhard de Chardin en Chine: Correspondance Inédite (1923–1940)*. Edited by Amélie Vialet and Arnaud Hurel. Aix-en-Provence: Edisud, 2004.

Teilhard de Chardin, Pierre. *Writings in Time of War*. Translated (from *Écrits du Temps de la Guerre [1916–1919]*, Paris: Grasset, 1965) by René Hague. New York: Harper & Row, 1967.

Teilhard de Chardin, Pierre, and Jean Boussac. *Lettres de Guerre Inédites*. Paris: OEIL, 1986.

Teilhard de Chardin, Pierre, and C. C. Young. "Preliminary Report on the Choukoutien Fossiliferous Deposit." *Bulletin of the Geological Society of China* 8 (1929), pp. 173–202.

Theunissen, Bert. *Eugène Dubois and the Ape-Man from Java: The History of the First "Missing Link" and Its Discoverer*. Dordrecht: Kluwer Academic, 1989.

Thorp, Holden. "Evolution's Bottom Line." *The New York Times*, May 12, 2006, p. A27.

Tobias, Phillip V. *Dart, Taung, and the Missing Link*. Johannesburg: Witwatersrand University Press, 1984.

Tobias, Phillip V. *Olduvai Gorge*, vol. 4. Cambridge, England: Cambridge University Press, 1991.

Tompkins, Jerry R., ed. *D-Day at Dayton: Reflections on the Scopes Trial*. Baton Rouge: Louisiana State University Press, 1965.

Trinkaus, Erik. *The Shanidar Neandertals*. New York: Academic Press, 1983.

Trinkaus, Erik, ed. *The Emergence of Modern Humans: Biocultural Adaptations in the Late Pleistocene*. Cambridge, England: Cambridge University Press, 1989.

Trinkaus, Erik, and Pat Shipman. *The Neandertals: Changing the Image of Mankind*. New York: Alfred A. Knopf, 1993.

Vanhaeren, Marian, et al., "Middle Paleolithic Shell Beads from Israel and Algeria." *Science* 312, no. 5781 (June 23, 2006), p. 1785.

Wade, Nicholas. *Before the Dawn: Recovering the Lost History of Our Ancestors*. New York: Penguin, 2006.

Walker, Alan, and Richard Leakey, eds. *The Nariokotome* Homo erectus *Skeleton*. Cambridge, MA: Harvard University Press, 1993.

Walker, Alan, and Pat Shipman. *The Wisdom of the Bones*. New York: Alfred A. Knopf, 1996.

Wang, Dominique. *À Pékin avec Teilhard de Chardin (1939–1946)*. Paris: Laffont, 1981.

Washburn, S. L., and P. C. Jay. *Perspectives on Human Evolution*. New York: Holt, Rinehart and Winston, 1968.

Washburn, S. L., and Elizabeth R. McCown, eds. *Human Evolution: Biosocial Perspectives*. Menlo Park, CA: Benjamin Cummings, 1978.

Weidenreich, Franz. *Apes, Giants, and Man*. Chicago: University of Chicago Press, 1946.

Weidenreich, Franz. "The Skull of *Sinanthropus pekinensis*: A Comparative Study on a Primitive Hominid Skull." *Palaeontologia Sinica*, new series D, No. 10 (1943), pp. 1–289.

Weiner, Steven, et al. "Evidence for the Use of Fire at Zhoukoudian." *Acta Anthropologica Sinica* 19 (2000), pp. 218–223.

White, Tim D., et al. "Assa Issie, Aramis and the Origin of *Australopithecus*." *Nature* 440 (April 13, 2006), pp. 883–889.

White, Tim D., et al. "*Australopithecus ramidus*, a New Species of Early Hominid from Aramis, Ethiopia." *Nature* 371 (September 22, 1994), pp. 306–312.

Williams, George C. *Adaptation and Natural Selection: A Critique of Some Current Evolutionary Thought*. Princeton, NJ: Princeton University Press, 1966.

Williams, George C. *Sex and Evolution*. Princeton, NJ: Princeton University Press, 1975.

Wilson, Edward O. *The Diversity of Life*. Cambridge, MA: Harvard University Press, 1992.

Zhu, R. X. "New Evidence on the Earliest Human Presence at High Northern Latitudes in Northeast Asia." *Nature* 431 (September 30, 2004), pp. 559–562.

Zollikofer, Christoph P. E., et al. "Virtual Cranial Reconstruction of *Sahelanthropus tchadensis*." *Nature* 434 (April 7, 2005), pp. 755–756.

INDEX

Harar Plateau, Ethiopia, 134
Hasebe, Kotondo, 200–201
Hastings Museum (England), 76
Heidelberg Man. *See Homo heidelbergensis*
Hicks, Herbert E., 114
Hicks, Sue K., 114
Hill, L. D., 111
Hirohito, Emperor of Japan, 201
Hironami, Nakada, 248
Hoang-Ho Pai-Ho Museum ("Licent's Museum"), 87, 90, 138
Hominidae. *See* Hominids
Hominids, 6, 14, 65, 155–57; Africa, 135, 186, 188, 229–32; Atapuerca, Spain, 56; australopithecine, 106–9; community life, 6, 156; evolution of, 43–44, 191–92; Java, 63, 189–90; and language, 236–37; Leakey finds, 230; Neanderthals, 47–49; Peking Man, 5–6, 173, 186; skull comparisons, 158–59; and stone tools, 45–46; timeline, 237; Zhoukoudian, 137, 141–43, 175. *See also* Anatomically modern humans
Hominization, Teilhard's ideas on, 221–22
Homo antecessor (Pioneer Man), 56
Homo erectus (Upright-Walking Man), 6, 56, 62, 150, 155, 159, 189–90, 229, 237; European discoveries, 232–33; first skull at Zhoukoudian, 5–6; and human evolution, 233–36; Java Man, 63, 68–70; language, 236–37; as missing link, 69; "Nariokotome Boy," 231–32. *See also* Java Man; Peking Man
Homo ergaster (Working Man), 159, 231, 234, 235, 237
Homo habilis (Able Man), 156, 188, 229, 230, 237; language, 236

Homo heidelbergensis (Heidelberg Man), 43–44, 56, 159, 237; as ancestor of modern Europeans, 234
Homo neanderthalensis (Neanderthal Man), 44, 47–49, 60, 62. *See also* Neanderthals
Homo rhodesiensis (Rhodesian Man), 106
Homo sapiens (Man the Thinker), 26, 58, 62, 237; Cro-Magnons as, 50–51; early, and symbolic thinking, 194; origins, 234, *234, 235*
Homo troglodytes (Cave-Dwelling Man), 26
Howells, William, 245
Huang Weiwen, 238
Human development, Teilhard and, 191–92
Human evolution, 46, 233–36; C. Darwin and, 105; Teilhard's view of, 6–7
Human habitation, in prehistoric Mongolia, 94–95, 95–98
Human races, Linnaeus's theory, 26
Humani generis (encyclical), 218
Humanity, Teilhard's ideas on, 194
Humans, modern: and Neanderthals, 58–59, 59–60; and Peking Man, 155. *See also* Anatomically modern humans
Humans, as species: C. Darwin's ideas on, 40–42; Linnaean classification, 25–26; origins of, 12–22, 186, 188, 229–30, 249–50; speciation, 221–22; Stone Age, 91, 92
Hunter, George, 111
Hunter-gatherer, Peking Man as, 156
Huxley, Thomas Henry, 33, 42–43

Ice Age, Neanderthals and, 48–49
Ignatius of Loyola, Saint, 6
Imperial Household Museum (Tokyo), 197
In His Image (Bryan), 115

Index of Forbidden Books, 75, 171
Inner Mongolia, 89, 91–93;
    Ordos Desert, 97
Institut Catholique (Paris), 76; Teilhard
    and, 82–84, 86, 99, 130, 208
Institute for Vertebrate Paleontology and
    Paleoanthropology (Beijing), 92
Intelligentsia, French, and Teilhard, 124
International Geological Congress (1933),
    172, 173–74
International Symposium on Early Man
    (1937), 190–91, 191
Israel: caves, 48, 55–56, 238; Neanderthal
    finds, 48–49; prehistoric humans,
    55–57
Italy, Neanderthal finds, 48

Janssens, Jean-Baptiste, 216–17
Janus, Christopher, 244–47
Japan: and Peking Man fossils, 7–8,
    200–201, 239–40, 240–41, 245–49;
    Teilhard in, 162, 193; war with
    China, 7–8, 192–93, 196–97, 203–4
Java: Dubois research, 68–69; Teilhard in,
    193. See also Java Man
Java Man, 7, 14, 63, 68–70, 156, 234;
    discovery site, 189; Japanese and,
    197; Peking Man and, 150
Jesuit authorities: and Teilhard, 6, 7, 22,
    79, 84–87, 99, 102–4, 121–24,
    130–31, 192, 199, 206, 208, 213,
    215–17, 220–25; and Teilhard's
    writings, 160, 195, 204
Jesuit order, Teilhard and, 73–80, 79–80,
    85–87, 102–4, 119, 121, 122–24
Jesuits: in China, 84–85; in France,
    73–74, 76, 77; in Peking, Japanese
    and, 204; Teilhard and, 101; and
    Vatican, 123. See also Jesuit
    authorities; Jesuit order

Jia Lanpo, 89, 140, 143–44, 238, 247
Johannesburg, 222
Johanson, Donald, 58, 109, 135, 155,
    230–31
John Paul II, Pope, 226

Kebara cave, Israel, 57
King Cheu-yen, 165–66
King, William, 47–48
Koenigswald, G. H. R. von, 69–70, 189,
    193, 197
Kolvenbach, Peter-Hans, 9
Krapina, Croatia, 48
Kromdraai, South Africa, 229
Kwangsi (Guangxi) Province, China,
    182
Kweiling (Guilin), China, 182

La Madeleine, France, 49
Laetoli, Tanzania, 231
Lamarck, Jean-Baptiste de, 22, 30
Lamare, Pierre, 133
Language development, 57–58;
    Cro-Magnons and, 50; genes for, 61;
    Neanderthals and, 57–58; roots of,
    238; symbolic thinking and, 236
Languages, Teilhard and, 90
Lapland, Linnaeus's studies in, 24
Larynx, as speech organ, 64
Lascaux cave, France, 50–51, 53, 224
Late Stone Age hominids, at Zhoukoudian,
    175
Lazarists, 152–53
Le Roy, édouard, 103, 124, 131, 171
Leakey, Jonathan, 186, 230
Leakey, Louis S. B., 105, 186, 188, 222,
    230
Leakey, Mary, 186, 230, 231, 232
Leakey, Richard, 186, 231; and language
    origins, 236–37

ABOUT THE AUTHOR

© Debra Gross Aczel

Amir D. Aczel is the author of fourteen books, including *The Riddle of the Compass, The Mystery of the Aleph*, and the international bestseller *Fermat's Last Theorem*. An internationally known writer of mathematics and science and a fellow of the John Simon Guggenheim Memorial Foundation, he lives near Boston.